Globalization, Urbanization, and Sustainability: What Can We Do?

Globalization, Urbanization, and Sustainability

What Can We Do?

First Edition

Abhishek Tiwari, Ph.D., M.P.H.

California State Polytechnic University – Pomona

Bassim Hamadeh, CEO and Publisher
Mieka Portier, Senior Acquisitions Editor
Skyler Van Valkenburgh, Project Editor
Casey Hands, Production Editor
Emely Villavicencio, Senior Graphic Designer
Kylie Bartolome, Licensing Specialist
Natalie Piccotti, Director of Marketing
Kassie Graves, Senior Vice President, Editorial

Contents

Introduction   vii

Chapter 1   Globalization, Urbanization, and Sustainability: The Three Big Ideas and How They are Linked   **1**

Chapter 2   The Environmental and Health Impacts of Civilization: Climate Change, Resource Consumption, Waste   **15**

Chapter 3   Urbanization and Globalization, a Historical Perspective   **45**

Chapter 4   Economic Fundamentals of the Modern Global-Capitalist System   **61**

Chapter 5   The Institutions of World Trade (IMF, World Bank, and WTO) and the Modern Financial System (Banks, Stocks, and Regulation)   **77**

Chapter 6   Demography: The Study of Population Characteristics and Their Relationship to Important Social Phenomena   **95**

Chapter 7   Measuring Poverty, Economic Development Approaches, and Poverty Policy   **113**

Chapter 8   Public Health: Tracking and Measuring Population Health and the Social Determinants of Health and Disparities in Health   **127**

Chapter 9   What Can You Do?: Why Should You Care, Imagining Alternatives, and the Domains of Action   **145**

References   **165**

Glossary   **185**

Introduction

This book is intended as an introduction to some parts of the big picture. But, what exactly is the big picture? Civilization has many parts that allow it to function and thrive. Though many would see technological growth as the defining feature of modern life, society's drivers are the changes in the underlying social system and its different parts (e.g., economics, political, cultural) that enable technological growth and the modern mind-set.

This system is, in many ways, incomprehensible; it is in some sense magical. The system produces a bewildering array of goods and services, whose production and workings we often scarcely understand. This is true of technologies such as computers, cholesterol medication or electricity infrastructure, systems such as finance, governance or air traffic control, or public services such as building permitting, epidemiology, or defense.

It is not that we are incapable of understanding, rather, our roles in society have become increasingly narrow. We are specialists, and depending on what we do, we understand at most one or a few pieces of the puzzle. Specialization has enabled complexity, technological growth, and the sheer abundance of goods and services that characterize our world today. By focusing on one or a few things, we are able to do that thing better, more quickly, and with greater precision. The benefits of specialization are self-evident: People live longer, have access to things and services that make life easier, and are less susceptible to the power of other people or their environment, including nature.

If this sounds like an overly optimistic or Panglossian view, it is. There are costs. Foucault, for example, argued that the assertion that humans are free from power is problematic as modernity is the increasing subjugation of humans to rationality. We have more stuff but are also more inextricably enmeshed in a multitude of systems for the provision of all goods and services. A large part of this book is intended as an introduction to those costs and their relationship to our modern systems of production. Sustainability is concerned with identifying and addressing these costs so that humans and the various flora, fauna, and nonliving parts of the planet's various ecologies also have an opportunity to thrive.

Human intelligence is responsible for these issues. It is the font of creativity and the source of its own destruction. Other creatures exist as part of nature. Humans instead seek to subvert, conquer, and outmaneuver nature. Human intelligence gave birth to complex civilization and allowed our population to grow to levels that would not occur naturally. However, human intelligence is unsatisfied to simply exist. It searches for meaning and creates competing narratives that sometimes conflict with each other. Human intelligence is also highly creative and seeks to understand itself, its environment, and the cosmos through arts and science. Intelligence compels humans to act in ways that lead to disequilibrium with nature. One could argue that this is inevitable as progress for humans is inextricably bound up with change and growth.

The modern world in many ways is a creature of enlightenment thinking, which includes a new epistemology based on verification of theories through observation, data collection, and analysis. This method is sometimes called the scientific method or positivism. Auguste Comte is often credited with formalizing the system used widely today to understand phenomena and cause and effect. Comte's positivism was political in nature and aimed at nothing less than a restructuring of society in the aftermath of the Enlightenment, which upended the older systems of meaning, namely those given by monarchies and the Church. Nietzsche's oeuvre was similarly aimed at addressing the problems, specifically the creeping nihilism that emanated from the death of older systems of meaning. Both Comte and Nietzsche sought to create new ways of thinking for a world evolving beyond the church and the king's authority. These new ways were needed to organize society, provide explanations, and confer meaning. Subsequent reformulations of positivism, sometimes termed postpositivism, emphasized its epistemology, or the process for "knowing things," over its sociopolitical aims. The scientific method today is thus more of a tool or a process about "how" to establish causal relationships and understand phenomena in the hard and social sciences than a philosophy that explores "why" things happen or a moral doctrine that provides the basis of human action. The scientific revolution enabled the modern age, a globalized, industrialized planet teeming with billions of people that produces a surfeit of goods and services for mass consumption.

Globalization—the word is evocative. To some it conjures images of exotic, far-flung places whose language, food, music, and art are made accessible by the networks, real and virtual, that tie the planet together. This sanguine view pictures globalization as a force that unifies and, ultimately, uplifts humankind. The economic efficiencies brought about by globalization allow societies to prosper materially and gain the freedoms to achieve their fullest potential. Trade engenders mutually beneficial interdependencies, which reduce the possibilities for conflict. Globalization is a rising tide that uplifts everyone. Others, however, see globalization mainly in pejorative terms. In this perspective, globalization is an economic force that brings the entire world into the capitalist fold, enriching the elites and their corporations at the expense of the masses who are further impoverished as their labor becomes increasingly cheaper. Furthermore, every aspect of their lives is made wholly dependent on the products and services provided by these corporations. The material benefits of globalization create a gilded cage as autonomy, self-reliance, and cultural singularity are sacrificed for material comforts. The cost is not only freedom but also the destruction of the natural environment.

These antipodal views are both correct. The role of globalization in reducing poverty, increasing the material quality of life, and impelling technological innovation is hard to overstate. The number and proportion living in extreme poverty today is considerably less than 3 decades ago. In 1990, over a third (35%) of humanity lived in extreme poverty.[1] Today, less than 10% exist in absolute penury. Additionally, the living standard and quality of life, as measured by access to goods and services, of all income groups is much higher than those in similar income strata in years past. The most dramatic differences are in the cost of technology, which has become widely available, cheaper, and considerably more powerful. For example, the Commodore 64, one of the first personal computers, cost $200 in 1984. This was roughly 1% of the median household income in the mid-1980s. Today, a basic laptop also costs about 1% of the median household income. However, the computing power of a modern laptop is greater by a magnitude of

---

1. The United Nations defines extreme poverty as "a condition characterized by severe deprivation of basic human needs, including food, safe drinking water, sanitation facilities, health, shelter, education and information." The current extreme poverty line is set at $2.15 per day.

hundreds of thousands.[2] Similarly, smartphones and other handheld devices with computing power superior to the fastest commercial computers 30 years ago are available to the masses in every country on the planet.

Another great technological creation is the internet. Within two decades of its creation, knowledge became democratized. Anyone with a connection to the World-Wide Web could access and create content on anything in a multitude of formats (print, audio, video). Ironically, this diluted the quality of information as anyone could create a web page filled with opinion masquerading as fact. Media content was no longer the purview of professionals. Social media allowed for the sharing and monetization of content created by the average person. Companies realizing the advertising potential of the internet developed algorithms to further steer users to others like them as verisimilitude of opinion ensured users would stay logged on and come back frequently.

The internet, instead of fostering dialogue, seems to have the opposite effect. It allows people to form communities with others across the planet with similar views and opinions. This creates echo chambers and reduces exposure to dissenting views, an essential ingredient of healthy and functioning democracies. The disagreements and hostilities around the pandemic and climate change show that science is also political: when cause and effect aren't clearly established, the internet gives voice to dissenting and minority opinions, which can be galvanized into a potent force. Today, we are at the cusp of an artificial intelligence (AI) revolution, which promises to upend society more than the internet.

Despite these technological improvements, extreme poverty persists, hunger and malnutrition are not a vestige of the past, and the chasm between the rich and poor has increased. Though living standards have improved across the planet, the income and the wealth of the rich have increased by even more. The rich have gotten incredibly rich. In the United States, for example, the top 20% of households today account for 51% of total household income; their share was 43% in the early 1960s when the Census first started tracking this metric. The top household quintile was the only group that increased its share of total household income in the last half century. This trend is not specific to the United States, as income and wealth[3] have become increasingly concentrated worldwide with the inequalities in wealth even more staggering. The ramifications are profound as concentrated income and wealth stymie upward economic mobility in even the most egalitarian nations by allowing those in higher SES groups, and their kin, to profit from economic rents,[4] further concentrating wealth, without directly contributing to the overall productivity of their societies.

The impacts on income and wealth inequality are problematic; however, globalization saves its wrath for the planet's ecosystems by increasing resource extraction, waste production, environmental contamination, and greenhouse gases (GHGs). Globalization favors urbanization as cities provide concentrated and skilled labor, multiple mar-

---

2. Computing power can be measured in terms of RAM, hard drive space, and processing speed. The Commodore 64 had 64 kilobytes of RAM and no hard drive. A basic laptop today has 4 gigabytes of RAM, over 1 terabyte of hard drive space, and a processing speed that is hundreds of thousand times faster.

3. Wealth refers to assets minus liabilities. Wealth includes assets such as stocks, bonds, real estate, precious metals, and so on. Income typically refers to wages earned in a time period. Income is typically taxed ,whereas wealth, as it is harder to estimate, is taxed when wealth is converted to income through the selling of assets such as stocks or real estate. The sale results in a capital gains event, which is taxed as income. Some countries have imposed wealth taxes. In the United States proposals for taxing wealth have been circulated, most notably be Senator Elizabeth Warren.

4. Economic rents are excessive returns or payments made to an owner of an asset that exceed the owner's opportunity cost. Rents are the excessive profits made by owners of an asset due to their positional advantage (e.g., when someone inherits stocks or real estate or someone lives off a trust).

kets, and economies of scale. Urban growth, however, often encroaches into the natural environment with the cost borne by flora and fauna, some of which perish permanently. Global finance requires companies to demonstrate regular growth or face stagnant or declining share prices, the *bête noire* of any large, publicly held corporation. Furthermore, companies are increasingly becoming part of the private equity sector, further removed from the regulatory oversight of publicly traded companies. More and more resources are extracted as there is a continual push to create new products and sell more; new customers are acquired, sometimes by pilfering from another company, or more is sold to the same customer base. For globalization to thrive, consumption must increase, but so does waste. The average American produces over four pounds of municipal solid waste (MSW) daily. Much of the waste is nonrecyclable and ends up in landfills or is discharged into the environment, including the oceans. Plastics account for the majority of the marine debris as plastics don't degrade quickly and are used widely in a variety of consumer products. Photodegradation breaks down plastics into microplastics, which are in our bodies and ingested by ocean life to their detriment.

Urbanization also exacts a heavy toll on human health. As people move to cities, they become more dependent on processed foods as they lack the time, skills, and space for cultivating their own food. The consumption of processed foods, particularly animal source foods (ASF), high in saturated fats, sugars, and other additives, is an ineluctable facet of increasing globalization and urbanization. The proportion of the population that is overweight or obese has steadily increased in every country in the last 50 years with a concomitant increase in diseases such as diabetes, heart disease, hypertension, and cancers. The factory farming of food has increased soil contamination and water pollution and is directly linked to an increase in greenhouse gases, specifically methane, which is released by cows raised for their milk and meat.

This book is neither a defense of globalization nor does it argue for its complete dismantling. Instead, this book seeks to help readers develop an understanding of the major drivers and impacts of globalization, particularly environmental consequences. Like most human phenomena, globalization is neither wholly good nor bad. Its impacts are context dependent, and it is shaped by a number of factors, some local and others global. It is not a panacea for economic underdevelopment, nor is it culpable for all that ails our planet. However, if we are to marshal its forces for a sustainable future, it is imperative to understand its relationship to various economic, social, cultural, political, environmental, and demographic contexts. We must also be able to imagine alternative futures, other ways of organizing society that go beyond the market, and we must each act, and act regularly, in ways that reflect our own unique skills, abilities and constraints.

This book is intended as a starting point for an exploration of this complex phenomenon. The first chapter provides an overview of urbanization, globalization, and sustainability and the ways they are connected. The second chapter describes globalization's impact on greenhouse gases and climate change and resources such as energy, water, and food and waste production. The third chapter traces the history of urbanization, specialization, and globalization by starting with the trade between two of the earliest civilizations in Mesopotamia and Indus Valley and ending with the world wars and the rise of the United States. The fourth chapter introduces the vocabulary and concepts necessary for understanding the economic foundations of globalization, specifically specialization (or comparative advantage) and free trade. Fundamental economic concepts, such as the market, competition, price, opportunity cost, absolute and comparative advantage, and externalities, are defined and elaborated upon. The fifth chapter extends the economic discussion to include national and international financial institutions, such as the Central Bank, World Bank, IMF, and WTO, and the monetary, fiscal, and trade polices enacted by these institutions to foster economic growth and trade. The sixth chapter explores the adage that "demography is destiny" by exploring the relationship between demographic concepts, such as fertility rate, dependency ratio and age structure, and economic growth. This chapter uses the economic and demographic vocabulary developed in the preceding sections to explore the reciprocal relationship between urbanization and globalization, helps clarify why people continue to flock to cities, and provides an overview of major urban challenges. The seventh chapter looks at the different ways of examining poverty, specifically the distinction between relative and absolute or extreme poverty, approaches to economic development internationally, and poverty policy in the United States. The eighth chapter focuses on the relationship between globalization and public health by providing the concepts necessary for understanding population health and examining the impact of globalization on human health and well-being, including the relationship between globalization and pandemics such as COVID-19. This chapter also explores social determinants of health and health inequalities, which are often exacerbated by the forces of globalization. The ninth chapter explores why we should care and act, the necessity of imagining different ways of being (ontologies) so that we seriously conceptualize alternatives to the existing world order, and the different domains of action from the personal to the macro (policy). The chapter revisits ideas developed in previous sections but also focuses on the importance of personal introspection, change, and action as the bedrock of larger systemic and global change.

Chapter 1

Globalization, Urbanization, and Sustainability

The Three Big Ideas and How They are Linked

Key Terms

Globalization: The economic, social, and cultural integration of the world. Economic globalization is the result of trade between regions and countries. Social and cultural integration is facilitated by trade, travel, migration, and the internet.

Urbanization: A process that results in the formation of cities. Characteristics of urban areas include large populations, higher populations densities, increasing dependency on markets for goods and services, diversity (cultural and social), greater resource consumption, and waste.

Sustainability: Typically defined in terms of three broad parameters (economic, environmental, and equity—called the three Es of sustainability). Economic sustainability includes poverty alleviation and development of economies that provide a high standard of living for all; environmental sustainability includes the use and recycling of resources to minimize pollution, waste, and externalities, or harms (externalities) to those outside the economic system; equity refers to societies that are nondiscriminatory, inclusive, and participatory across various dimensions such as gender, sexual orientation, age, race, and so forth.

Other key terms will be in **bold** and defined in the text.

Review Questions

- Why do urban areas form? What are the benefits of urbanization? What are the problems associated with urbanization?
- What is specialization, what are its benefits, and how is specialization linked to trade, urbanization, and globalization?
- What are markets, and why are they important?

- What are economies of scale? What is economic efficiency?
- What is the role of specialization and assembly line production in creating efficiencies and economies of scale?
- What are the different ways that urbanization, globalization, and sustainability interact?
- How is sustainability conceptualized? What are its dimensions and the tensions between them?

Introduction

Globalization, urbanization, and sustainability are intimately connected. **Urbanization** enables division of labor and specialization, clustering of people and industries, and the development of real or virtual **markets,** which bring together buyers and sellers of goods and services. These factors are integral for economic globalization. The enormous amounts of goods and services produced in a global-capitalist economy require substantial inputs in terms of resources and labor. Increasing production also leads to incredible amounts of waste. Thus, both resource extraction and waste production increase significantly as regions industrialize and become more integrated into the global economy. Furthermore, the fruits of economic growth are often unevenly shared, with the greatest spoils often going to those who control capital, the means of production and money. Sustainability-related measures seek to redress these development issues by balancing economic development, environmental impacts, and equity-related issues.

The first urban areas or city-states arose in Mesopotamia, what is today Iraq, over 5,000 years ago. Agricultural innovations, such as crop domestication, enabled a food surplus so people could specialize in nonagricultural activities such as the production of pottery, weapons and clothes, or vocations related to specialized skills such as writing (scribes), politics (government), or religion (priests). [1] This coincided with the development of actual markets in urban settings for the exchange of goods and services and long-distance trade. The factors that helped foster the formation of the first cities in Mesopotamia were thus instrumental in ushering in the first era of economic globalization.

Specialization, or focus on a particular skill or product, and trade led to the growth of cities, then kingdoms and eventually empires. These rose and fell in the ensuing millennia with some, such as the Romans, managing to sustain themselves for centuries. This pattern would continue until the confluence of two factors: industrialization and American independence. The industrial revolution vastly increased the speed and quality of goods production, and the American experiment created the world's first modern nation-state based on the consent of the governed (i.e., democracy). The political equality promised by the American revolution was complemented by the relatively greater material equality created by the industrial revolution, which vastly increased the goods available for consumption for all strata of society. The 20th century would witness an explosion in human population, as advances in public health reduced mortality and substantially increased the material quality of life and warfare on an unprecedented scale; also, environmental degradation and climate change became major ongoing issues. The same factors that increased the human population and quality of life also, ironically, were responsible for the biggest era of destruction on a planetary scale.

---

1. This view is simplistic as the movement towards agriculture and specialization was not linear. Many cultures experimented with a number of food procurement strategies such as foraging, hunting or agriculture and specialization certainly existed prior to large scale agriculture. Nonetheless, these trends accelerated with the adoption of agriculture and the formation of cities.

This chapter explores the economic basis of globalization, specifically its relationship to markets, specialization and the Fordist economy. Additionally, the chapter looks at theories of urbanization, or why cities form, the effect of urbanization, and the peculiarities of suburbanization in the United States. The chapter concludes by providing an overview of the environmental movements in the United States and the three dimensions—economy, equity, and environment—used to frame sustainability today.

What Is Globalization?

Globalization implies connection, exchange, and interdependency.[2] Historically this has meant networks of trade between regions through which people exchanged goods, services, and ideas. Interdependency emerged as these networks matured, and people in the regions engaged in trade came to rely on goods produced elsewhere. Today, interdependency means that the **supply chains**, or various producers for different components of a finished product, for many of the goods we rely on daily are scattered across the world. Information and services have become important commodities as virtual exchange facilitated by the internet has reduced the transportation time for receiving certain goods and services to almost nothing. The internet has also enabled an unprecedented exchange of ideas and culture as the various forms of social media have become an indelible part of daily life. Markets and **market dependency** have become the defining feature of modernity as individuals across different societies specialize and utilize different markets for almost all aspects of their daily existence. The wholesale export of these **ontologies**, or ways of being (i.e., democracy, markets, capitalism, individuality), mean that even those engaged in **subsistence agriculture** (those who make their own food) are brought increasingly within the folds of the market and specialize in the specific crops for which they have a **comparative advantage** (i.e., those crops they can specialize and grow best based on their skills, resources, and climate). Development theories are overwhelmingly market based. The earliest versions, like those based on the Washington Consensus, often called for structural reforms to create competitive markets that would initially plunge poorer countries into a period of "necessary" economic pain that would eventually orient them toward a trajectory of long-term growth. When this failed to occur in developing regions in Latin America, Africa, and Asia, and China lifted several hundred million out of extreme poverty without democracy or structural reforms of the type advocated by the International Monetary Fund (IMF), the recipe changed and included more protections for indigenous producers and a bigger social safety net; however, the primacy of the central tenets of global capitalism, namely urbanization, specialization, and markets, remained. There have been more hiccups along the way: the 2008 global recession (or a depression, depending on where you were and what jobs you were engaged in); the growth of far right parties and policies in the second decade of the 21st century; the COVID-19 pandemic and resulting backlash against government interventions; and increasing climate change. To its proponents, these are temporary setbacks that can be managed within the framework of global capitalism; to others, the problems are caused by the various ideologies of global capitalism.

2. The International Monetary Fund (IMF) defines globalization as "a historical process, the result of human innovation and technological progress, which leads to the increasing integration of economies around the world, particularly through the movement of goods, services, and capital across borders."

The Growth of the Fordist Economy

At the heart of the current global capitalist regime lies the very simple idea that the modern world, with its many marvels, comforts, and appurtenances, is impossible without specialization, which, when properly harnessed and organized, allows for **economies of scale**—*advantages that come from size*—that improve production efficiency and product quality. Economies of scale increase product quantity and quality, the sale of which provide the capital needed for further research and innovation. Capital is also raised from the sale of equity, or company stock, on private or public markets. Cheaper and higher quality consumer products increase a society's material standard of living. In the end, at least theoretically, everyone is better off. **Panglossian**, or overly optimistic view, however doesn't address environmental issues resulting from mass production, including climate change and waste. Before delving into these nuances, a brief history of Fordist revolution is instructive, and what better place to start than its eponym?

In 1913, Henry Ford, hoping to bring the automobile (Model T) to the masses, put the idea of mass production and economies of scale into practice by introducing the first industrial assembly line. By dividing the functions needed to assemble the finished product into smaller, repeatable tasks in a smoothly moving assembly line, Ford was able to achieve greater production efficiencies and consistency of quality. Within 10 years, the Model-T price dropped from \$950 to \$290 in 1924 (2). Initially, manufacturing plants were hemorrhaging workers as the work was tedious. To reduce worker attrition, Ford increased daily wages to \$5, which was nearly double the wages paid in other plants (2). The workday was also eventually reduced from 9 to 8 hours to accommodate three daily shifts. By paying workers more, Ford increased worker retention and expanded the customer base for his automobiles as the average car price was now equivalent to roughly 3 months of gross wages.

Ford had first observed the production efficiencies gained by the movement of goods on conveyor belts in slaughterhouses and granaries. Working with Fredrick Taylor, one of the fathers of the scientific management system, Ford created a system that broke down the manufacturing process for the Model T into over 80 smaller steps that all occurred on a moving assembly line. An important component of the Fordist production process was the use of standardized components, an uncommon practice in the auto industry at that time. By using the same parts in every automobile and training workers for specific tasks, Ford was able to greatly streamline workflow. This increased worker efficiency and ultimately industrial output. Ford was soon producing more automobiles than all other auto manufacturers combined. The lessons of assembly line production were applied to the war machine during WWII as the United States became the primary producer of military hardware as European manufacturing was decimated by war (3). At the height of the war effort, for example, the United States was producing a plane every 63 minutes (3). The manufacturing lessons learned during the war would be instrumental in cementing the domination of American manufacturing through the mid 20th century.

The Market

In his seminal book, *Wealth of Nations*, Adam Smith, considered by many to the father of modern economics, argued for the primacy of markets in advancing the material prosperity of nations (5). Coincidentally, the book was published in 1776 and undoubtedly influenced many of the young nation's early leaders. In a 1790 missive, Thomas Jefferson noted that *"in political economy I think Smith's wealth of nations the best book extant"* (1). Smith believed that markets are best suited to actualize the creative and competitive impulses that are inherent in each individual. Smith did not envision a society free of all government. Rather, the government's role was best suited to enterprises such as defense or education (5). Furthermore, the judiciary was important in providing the legal framework necessary for sustaining free and fair markets. Smith's thinking was elaborated by many, including David Ricardo, who first articulated the idea of **comparative advantage**, which asserts that regions should specialize and produce those products that they can produce most efficiently or have the lowest opportunity cost (this concept will be explored in Chapter 4) (4). Increasing specialization will increase the volume and quality of available products, which can be exchanged through networks of trade, thereby increasing the material standard of living for all.

In classical liberal and neo-liberal thought, markets, as opposed to governments or a central planning apparatus, are best suited to transform labor and material inputs into productive ends or products. In an ideal economic system, resources are always directed toward ends with the greatest utility for society. **Utility**, a concept that includes happiness, in traditional microeconomic theory, is maximized by individuals through choices, namely consumption preferences, made in a smoothly functioning market (6). **Markets**, replete with buyers and sellers and free from **distortions**, or factors such as price manipulation that skew market functioning, are thus best suited to actualizing the wants and desires of the myriad souls inhabiting this planet. Friedrich Hayek provided the intellectual basis for the primacy of markets by arguing that no one individual or group can possibly be privy to all relevant facts as knowledge is "fragmented and dispersed." The market is thus best suited to creating a "spontaneous order" that best reflects the choices made by the multitudes engaged in the market. This view of markets, though decidedly Panglossian, or overly optimistic, soon gained adherents throughout the world. By the 1800s it was the organizing principle for industrializing countries; factory production and markets were the way of the future. The events of the early 20th century would expose the shortcomings of this view. The Great Depression waylaid the seemingly inexorable rise of markets and **supply-side economics**, which argues for the central role of suppliers or producers in the economy in providing goods and setting prices to which consumers respond (7). The Great Depression gave birth to demand-side economics, which argued government policies that stimulated aggregate consumer demand, and not producers, were the prime movers of the economy. Whether one subscribed to supply-side or demand-side view, the mechanisms through which society moved forward were still very much in the realm of markets.

John Maynard Keynes, considered the founder of **demand-side economic theory**, argued that aggregate demand, or total demand for goods and services, was the primary driver of the economy (7). Demand-side economists argue that when people and the private sector are unable or unwilling to provide the requisite demand to fuel economic growth, the government, through various fiscal and monetary actions, can and should provide the necessary stimuli to jumpstart aggregate demand. This belief in the power of the government to stimulate aggregate demand was the foundation of the New Deal, passed in response to the Great Depression, as well as the American Recovery and Reinvestment Act, passed in response to the 2007–2008 recession. The various COVID-19-related relief bills by both the Trump and Biden administrations in response to pandemic-induced economic havoc are also examples of demand-side stimulus spending. Active and ongoing interventions in markets are contrary to classical liberal economic theory as well as the various strands of neoliberal thought based on the work by Friedrich Hayek and his protege, Milton Friedman. Hayek, as mentioned earlier, argued this perspective most forcefully and cogently. In inextricably linking individual freedom with markets, Hayek laid the foundation for generations of libertarians and their ilk. Hayek was critical of centralized economic planning and the Keynesian welfare state, the latter which he later came to associate with the march toward totalitarianism. Early development theorists embraced Hayek's view as well as Keynes's prescriptions for lifting poorer countries out of poverty.

Development economics emerged as a discipline in the aftermath of the Great Depression and World War II. The premise was simple: The political economy of poor nations could be engineered to raise living standards over time by providing more and higher quality goods and services to the populace. The Bretton Woods conference was convened in 1944 to create a financial system to ensure global economic security. The conference engendered the International Monetary Fund (IMF) and the World Bank, the two major institutions that, along with the World Trade Organization (WTO), continue to manage the world economy and trade today.

The rejoinders to those who valorize market-based, global capitalism and related development theories are many. **Marxian analysis**, for example, focuses on the increasing alienation and subjugation of the worker and increasing income and wealth inequality under a capitalist regime (8). Marxian and related critiques generally focus on the idea that the owners of the means of production (i.e., the capitalists) are able to increasingly usurp the productive capacities of the worker. The surplus value generated by the worker accrues disproportionately to the capitalist who captures the excess profit if the market price (i.e., actual price charged) for their good is above the natural price (i.e., price that is based on actual costs of inputs or costs of production) (8). Over time the number of capitalists become fewer (i.e., wealth becomes concentrated) and the laborers many. So, over time, the concentration of capital increases the division of labor, and larger numbers become dependent on fewer capitalists for their means of subsistence (i.e., wages). This process enables the capitalist to exert a constant downward pressure on worker wages. The capitalist, because they have access to capital, is also able to change domains (i.e., area of production) easily, whereas the worker having spent an entire life in one domain is a prisoner to their domain-specific skill. Rising worker incomes throughout the 19th and 20th centuries, the creation of the middle class, and the general economic prosperity in industrialized countries during the 20th century, undercut Marxian critiques. However, today wealth and income inequality are severe in many developed countries, including the United States, and show no signs of reversing. Marx may yet see his arguments come to fruition if these trends are not addressed by both the private and public sectors.

Another important critique comes from **environmental economics,** which argues that the traditional economics perspective on resources is flawed as externalities and environmental value are ignored (9). In traditional economics, resources should be allocated to those users who are willing to bid the highest price for that resource. The higher prices reflect consumer demand and utility. However, in such a view there is no economic value for leaving resources unused unless an entity is willing to pay a price to "not use" the resource. For example, the best use of a tree would be for the industries, such as paper or construction, wiling to pay most for that tree. In the market for trees, the trees would remain untouched only if the price bid by the conservationist exceeded those of the paper or construction industry.

These are not the only critiques. Other scholars have looked at markets and capitalism from other perspectives (e.g., subaltern, postcolonial, gender, postmodern, etc.). There are many cogent arguments for why markets and capitalism are problematic. Some frame the issue epistemologically as part of the increasing rationalization of human life that started with the enlightenment period, which gave birth to the scientific method. Others see it as a Faustian bargain from which there is no escape as the comforts provided by the current system make effective or radical critique impossible. Whatever the case, unless we collectively believe that different ontologies (i.e., ways of being and living) are both possible and desirable, we will forever be trying to put a better bandage on the problems instead of thinking outside the proverbial box.

Conceptualizing Urbanization: Why Do Cities Form, and What Are Their Impacts?

The seemingly inexorable process of urbanization in the context of globalization, which has resulted in more than half the population living in urban centers, exerts a homogenizing influence on cultures and peoples that, intuition suggests, will lead to a common cosmopolitan mélange in regions throughout the world. More than half of the planet's population already resides in an urban area. The proportion in cities is only expected to grow in the coming decades. As the world's population stabilizes, more than three fourths will be urbanites. The largest of these areas will be mega-cities with 10s of millions living within city boundaries. Some of these **alpha** cities already exist today; others will arise in developing regions (10). From Los Angeles to Sao Paolo, Shanghai to London, or Johannesburg to Sydney, the modern city, its many appurtenances, facets, and dynamics, seems almost a foregone conclusion: the teleological result of economic and social processes that could not have resulted in an urban form other than what is observed in major cities the world over. In other words, the city and its characteristics seem inevitable, almost natural, and what came before was primitive, a prelude to the wonders of the modern, market-based, techno-capitalist, hyperconnected world.

The city as the dominant form for organizing man's social, political, economic, and cultural life begs the question "Why?" Specifically, why do cities arise, and how do they impact their residents? Scholars from various disciplines have theorized the city to discern the forces that create them as well as their impact on those that live there. Economists see cities in terms of their potential for increasing economic productivity by clustering both people and industry to achieve economies of scale. Cities contain many different types of markets in addition to job markets (e.g., a marriage market) and, based on their location, specialize in certain types of production and engage in specific types of trade (city location is thus essential for understanding the comparative advantage of different cities) (12). From an economic perspective, cities also have location tradeoffs as people must negotiate the benefits of income and amenities with the costs of housing and transportation (11). In other words, some areas in a city may offer greater amenities, or proximity to work, but these are offset by, for example, higher housing costs. **Economic theories** assess urbanism in terms of the changes in political economy and the productive efficiencies resulting from these changes. **Sociological theories** look at the changes in human relationships in urban areas and the implications of these shifts for human well-being. Marxist-oriented scholarship, as discussed earlier, bridges these two disciplines by looking at the impact of capitalist modes of production on social outcomes, namely the concentration of capital and the concomitant social costs.

The various sociological perspectives offer unique insights as they weave together ideas from different academic corners to theorize the types of human relations engendered by the city; specifically, the sociological theories attempt to shed light on the social dimensions of city life that make it distinct from other types of human habitations and the particular or peculiar psychologies that arise in urban life. George Simmel was one of the first sociologists to dissect urban life. His analyses mainly centered on the impact of the money economy on human psychology. Simmel argued that the urban economy and intense sensory inputs of city life compel man to become more precise, calculating, and blasé (13). Additionally, all values are reduced to a monetary equivalent. In other words, money becomes the sole repository of value as the different values inherent in different things are transmuted into a monetary value: Everything ultimately is for sale! In *The Division of Labor in Society*, Emile Durkheim, often credited as one of the chief architects of sociology, discussed the evolution of society, specifically societal cohesion, in terms of the transition from mechanical solidarity to organic solidarity (14). In simpler societies, such as those characterized by subsistence agriculture, individuals bond because they are similar in many respects. This is **mechanical solidarity**. Individuals engage in similar types of work; they have similar upbringings and hold similar worldviews. Though rudimentary trade and bartering exist in such societies, the individuals in these societies are relatively undifferentiated or homogenous and can produce nearly all the things they need themselves. As societies become complex, individuals specialize and become interdependent, as no one individual can produce all things needed for survival. This type of society is highly differentiated and characterized by **organic solidarity**. Ferdinand Tönnies, also seen as a founding figure in sociology, described the shift from rural to urban in terms of the transition between two ideal types, **Gemeinschaft** (community) and **Gesellschaft** (society) (15).

The former has mainly an agricultural economy and is organized around small groups living in a village or small town. The latter is based on an industrial economy and trade that occurs primarily in an urban setting. Tönnies, like his fellow sociologists, was interested in the psychological orientations associated with urban and nonurban life. He posited that *Gemeinschaft* relationships are sentimental whereas those in *Gesellschaft* are instrumental, or practical (15). Like Simmel, Tönnies saw humans as becoming more rational, calculating, practical, and self-interested with the shift to urban living. Louis Wirth, who provided perhaps the most comprehensive early conceptualization of urbanism, defined a city as "a relatively large, dense, and permanent settlement of socially heterogenous individuals." Like his predecessors, Wirth argued that urbanism makes individuals rational, calculating, and primarily self-interested. This is both liberating, as the individual is no longer susceptible to the constraints sometimes imposed by intimate or sentimental relationships, and stultifying as relationships become transitory, superficial, and instrumental. The result is **anomie**, or normlessness, a concept first used by Durkheim to describe the social basis for the dysfunction he observed in complex, urban societies (14). Basically, in this view, cities made individuals more interdependent but also more isolated and ultimately more dysfunctional. Later sociologists, from compositional and subcultural schools, would reject the link between urbanism and social disintegration. Instead, they argued, cities allow the formation of smaller social groups based on class, ethnicity, age, or some other compositional element. These groups have the same kinship and filial relationships as villages and towns, though they are ensconced within large urban frameworks. Additionally, subcultures, especially those viewed as taboo or deviant in smaller communities, have spaces to thrive in large urban areas. Urbanism, instead of weakening human bonds, can create new bonds based on a shared subculture and strengthen existing ones based on a compositional element such as age, ethnicity, or economic class.

Suburbanization in the United States

By the end of the 1st decade of the 20th century, the United States was already a nation of car aficionados: There were 24 companies producing automobiles in 1908. In 1920, more than half of American families owned at least one automobile, and one in five had more than one (3). During this time, the cars were many, but the roads few. The ethos that roads belonged to the automobile had not yet fully formed. Roads were populated by people, bicycles, horse-drawn carriages, and trolleys. Roads were also open space and used for recreational purposes by all ages. However, as automobiles increased, others uses were relegated to the periphery or, as in the case of the horse-drawn carriages, became obsolete. By the end of the 1920s, the primary function of roads was automobile thoroughfares. Roads gave birth to another phenomenon that would contribute heavily to many of the environmental and health problems we face today in the United States: the suburbs. (16).

An urban core ringed by suburban settlements is a recurring leitmotif of urbanization throughout the world. **Suburbs** have been around as long as the first urban areas in the fertile crescent. As the first city-states developed, communities developed around the city boundaries demarcated by walls. Similarly, the urban core of Mayan cities was primarily used for administrative and ceremonial purposes, with much of the population living in settlements around the core (17). European cities developed along similar patterns. The word *suburb* itself was likely first coined sometime in the Middle Ages. Though suburbs have existed in various forms for several millennia, the modern suburb is a creature of industrialization that first occurred in London and then in other major cities throughout Europe and the United States. The suburbs became a way to escape the pollution, crime, congestion, and other ills associated with rapidly industrializing cities. For example, the Garden City movement, popularized by Ebenezer Howard in 1898, sought to balance the benefits of nature and urbanism (18). Suburbs became a way of combining the amenities of the city with the open space of rural areas.

Today, the United States is primarily a nation of suburbs. It's important to note the delineation between urban, suburban, and rural is not based on some fixed criteria but dependent on the features (e.g., open space, density, land use, etc.) one uses to organize areas. The U.S. Census director suggested that there is a growing consensus in the demographic community that "statistical quality" is measured by how "fit" a particular statistic is for a given use. This is not to say that the categories are meaningless, but rather there are different ways to operationalize the term. These limitations notwithstanding, the trend toward suburbanization is clear. The 2017 American Housing Survey (AHS) shows most Americans (52%) identify their neighborhood as suburban, with 27% indicating they live in an urban area, and another 21% indicating rural residency (19). These results were similar to those generated by a 2015 survey by Trulia, an online real estate–focused company.

Suburban neighborhoods in the United States have several characteristics that set them apart from urban or rural areas. In general, suburbs tend to be characterized by single-family homes, higher median household incomes, and a college-educated population. Suburbs also have a larger proportion of non-Hispanic Whites, homeowners, married households, and households with children. The demographic characteristics of U.S. suburbs mirror how suburbs are seen in the popular imagination.

Suburbanization is the result of the interplay between several economic, social, political, and technical factors. The processes that drive suburban formation in the United States include **demand–side or Keynesian economic policies**, such as tax credits for buying EV vehicles, which are implemented by the government to stimulate consumer spending, economies of scale, the extension of credit for various goods and services, government intervention in credit markets, rising household incomes, availability of wood as a building material, the privacy ethos, government-enabled infrastructure development, planning regulations, and general improvements in the material quality of life brought about by these factors (which will be discussed in greater detail in Chapters 3 and 4). The implications of suburbanization for the economy and the environment have been significant. Suburban growth is an important driver of the economy. Consumer spending increases as homeowners use their home equity, which increases over time due to price appreciation, to finance consumer purchases. Homes also exert a multiplier effect on local economies through their impact on related industries such as construction and real estate. The impact of suburbanization on the environment is substantial. Suburban growth results in lower density development, which requires greater resource consumption. Water and energy use as well as waste production are much higher in suburban areas than higher density areas with comparable populations. Lower density developments also increase transportation distances for all types of trips (i.e., work, recreation, school, consumer related, etc.), increasing both energy used for travel and climate change gases resulting from increased travel. Additionally, traffic-related congestion exacts a heavy toll on mental health and costs the economy billions in lost productivity.

Sustainability: A Brief History

The environmental movement in the United States can trace its history to well-known public figures such as Henry Thoreau, John Muir and Theodore Roosevelt as well as lesser known but equally important individuals such as historian and naturalist George Grinnell and Gifford Pinchot, the first head of the U.S. Forest Service. There have been several environmental movements in the United States since the mid 1800s. These include preservation (1850s), conservation (1890s), public health (1920s), environmental protection (1960s), environmental justice (1980s), sustainability, and climate change (1990s onward) (20). The various movements, though similar in some ways, have notable differences. The **preservationists** like Thoreau and Muir were interested in preserving nature for its own sake, whereas the **conservationists** during the late 19th and early 20th century were primarily interested in the sustainable use of natural resources, which included both flora and fauna (20). The ideas of reforestation and sustainability yields came of age during this era. The **public health movement** helped link the emerging understanding of communicable diseases and pathogens to the environment, specifically cleaner air and water (20). Sanitation and the treatment of potable water were a major theme of this movement. The 1960s were a time of social and political upheaval in the United States. The modern environmental movement was also born during this decade.

Spurred by the publication of Silent Spring by Rachel Carson, who highlighted the deleterious impacts of pesticides, and major environmental disasters such as the Santa Barbara oil spill and the Cuyahoga River fire, the movement led to the passage of the landmark National Environmental Protection Act (NEPA) and the birth of the Environmental Protection Agency (EPA), which was signed into existence by Richard Nixon. Other major environmental legislation would follow and included the Clean Air Act (1970), the Clean Water Act (1972), the Safe Drinking Water Act (1974), the Toxic Substances Control Act (1976), and the CERCLA Act (1980) which created the Superfund program for environmental remediation (20). The **environmental justice moment** of the 1980s focused on income and race- (ethnic)-based disparities in exposure to environmental contaminants and resulting health outcomes. **Climate change**, which refers to manmade or anthropogenic global warming that leads to extreme weather and other environmental outcomes, became a major theme during the 1990s with the establishment of the Intergovernmental Panel on Climate Change (IPCC), which recommended the formation of the UNFCCC in its first assessment report (21). The UNFCCC convenes the conference of parties (COP), a meeting of world governments to address issues related to the environment, climate change, and sustainability. The COP meetings have led to some important international agreements around sustainability, including the Kyoto Protocol in 1997, the first international climate protection agreement; the Paris Accords in 2015, the first legally binding international treaty on climate change; and the joint declaration by United States and China in 2021 during COP 26 to work together on a number of climate change–related issues to meet the goals of the 2015 Paris Accords (22).

The environmental movement is large and complex, as attested to by its history in the United States, with many different, sometimes competing, agendas. For example, a perspective shaped by conservation of nature for sustainable use by humans leads to different policies than a perspective driven by preservation of nature for its own sake. Today, sustainable development is typically conceptualized within the context of three domains: economic, environment, and equity (sometimes referred to as the three E's of sustainability). Economic concerns include poverty reduction, specifically the elimination of extreme poverty as measured by the proportion of those living on less than $2.19 per day (in 2020 dollars); income and wealth distribution; and regulation of capital and finance markets to ensure global stability and fairness. Environmental concerns include climate change, resource depletion, ecosystem and habitat destruction, waste-related impacts, and maintenance of biodiversity and natural resources for their own sake and not economic uses. Equity-related issues include gender equality; treatment of marginalized groups; political freedoms, participation, and accountability; human rights abuses, education access, and a social safety net. These ideas were articulated by the United Nations into a set of sustainable development goals that cut across many different sectors of society and seek to balance the tensions between economic development, environmental impacts, and equity-related concerns.

The United Nation's Sustainable Development Goals

On September 25, 2015, the United Nations adopted 17 **Sustainable Development Goals (SDGs)** to end poverty, eliminate hunger, promote good health and well-being, provide quality education, increase gender equality, ensure the availability of clean water and sanitation, foster the development of affordable, clean energy, promote economic growth and decent work, develop industry and infrastructure, reduce inequalities of all types, promote sustainable urban environments, inculcate responsible consumption and production, encourage action on climate change, protect ocean and land life, foster strong institutions that ensure justice and peace, and develop partnerships to achieve these goals (23). The SDGs were an expansion of the eight Millennium Development Goals (MDGs) adopted in September 2000. The MDGs focused on poverty, hunger, education, gender equality, child and maternal health, communicable diseases, including HIV, and environmental sustainability. Between 2000 and 2015, many countries made significant progress toward achieving the MDG goals (23). An important victory was the significant increase in official development assistance (ODA), or aid directed toward developing regions: In 2014, $135 billion was provided as ODA (24). Other major accomplishments include more than a 50% reduction in absolute poverty (i.e., those living on less than a dollar a day, or $2.19 in 2020 prices) from 1.9 billion in 1990 to 836 million in 2015; a decrease in the prevalence of undernourishment in developing regions from 23% to 13%; an increase in primary school enrollment in developing regions (from 83% in 2000 to 91% in 2015); exceeding the targets for reducing incidence of HIV, TB, and malaria; and decreasing child and maternal mortality (24).

The SDGs reflected a more sophisticated understanding of the relationship between the various **dimensions of sustainability**. Specifically, the goals were created with the understanding that economic, equity, and environmental concerns intersect in a multitude of ways with reciprocal impacts. The SDGs also reflected a desire to strengthen the science-policy interface (SPI) for sustainable development. To this end, the UN developed a network model to illustrate the quality and quantity of connections between the 17 areas based on expert (scientific) assessments (23). Today, the SDGs provide the framework and targets for worldwide action on the various dimensions of sustainability. The most recent status report on the SDGs notes that though progress has been made in some areas such as material and child health, infrastructure access, and women's representation in government, the COVID-19 pandemic will reverse some of the gains of the past two decades and push tens of millions back into extreme poverty, increase economic vulnerability among those in the informal economy, increase health issues among the poor, particularly those living in slums, reduce educational opportunities for poor children, and increase domestic violence (25).

Globalization, Is It Good or Bad?

Globalization's boosters come from all parts of the political spectrum. It inspires support from those on the right who argue for market-based capitalism and little regulation as well as those on the left who champion democracy, human rights and other, related, liberal values. Some see the march of globalization in terms of teleology, an ineluctable process that results in human emancipation, both economic and political. The great project of Western liberalism is in many ways the marriage of democracy to free market capitalism. The last bastions against democracy and capitalism fell during the 1970s through the early 1990s: China opened up its economy, the Berlin Wall came crashing down, and the USSR and its satrapies dissolved almost overnight. These events were seen by many scholars as harbingers of a new and better world older, one characterized by the ascendancy of Western liberal values. However, some saw these optimistic prognostications as naïve or overly sanguine. They were validated by the events of the first two decades of the 21st century. These included the great recession in 2007–2008 and the election of public officials that were overtly antiglobalist and deftly used identity politics to divide, exclude, and demonize. The Trump administration in the United States, for example, sought public and private funds to build a wall to stem illegal immigration, halted Muslim immigration until officials "could figure out what was going on," and vowed to bring American jobs back (26). The Trump administration also pursued an "America First" policy in international relations by revising or repealing many of the earlier economic and other polices it viewed as overly conciliatory (26). The most bombastic rhetoric was, not surprisingly, directed at China, America's biggest trading partner and geopolitical competitor. However, even American allies, especially those in Europe, were not spared as the Trump administration promised to make them "pay their fair share" for NATO and other "policing the world"–type activities (26). Similarly, the Brexit vote in Britain was seen by many as yet another nail in the globalization coffin. The events of the past four decades have demonstrated time and again that globalization, like much of human behavior, does not fit neatly into one or even a few theoretical frameworks. Similarly, capitalism has not evolved uniformly in the countries that came to adopt markets in the last few decades. Chinese capitalism is very much state controlled, the Russian version can loosely be called a kleptocracy, and the U.S. version continues to privilege corporations at the cost of just about everything else (27). Yet despite these differences, and the continuing geopolitical jousting, the United States and China, in principle at least, agreed on a joint framework to address climate change at the UN's COPT 26 conference in 2021.

In addition to increasing the material standard of living, globalization is integral for the intellectual development of humanity as it increases the exposure of individuals and groups to each other. The **real and virtual networks** between regions that enable trade also foster the sharing of ideas, information, and culture. The explosion of internet-based platforms during the last 15 years for sharing content that is loosely grouped under the banner of social media attests to the important role of globalization in promoting "idea-diversity." Just as **biodiversity**, or the existence of variety of plants (**flora**) and animals (**fauna**) in an area is essential for the health of an ecosystem, **idea diversity**, fostered by exposure to different perspectives and ways of living, allows people to learn, evolve, adapt, and thrive. History has repeatedly shown that most insular societies eventually face an existential crisis as they lack the capacity to evolve, change, or adapt.

The historical record is filled with the detritus of empires and people who once ruled mightily only to succumb to their hubris, the belief that they knew and understood better than anyone else. Even the most successful conquerors recognized the importance of engaging and assimilating the ideas of the conquered. Alexander the Great, for example, had a habit of adopting local customs. Similarly, Genghis Khan was thoroughly ecumenical when it came to the various religions under his dominion. Religious intolerance was verboten under Mongol law. The spirit of this law animated no less a person than Thomas Jefferson, who had read a best-selling 18th-century biography of Genghis Khan (28). Perhaps it is not that much of a stretch to argue that the much vaunted First Amendment of the U.S. Constitution is a legatee of the Great Khan's strictures against religious intolerance.

Despite its many positive contributions to the material and intellectual development of humankind, globalization exacts a heavy environmental toll. The basic structure of market-based economies is, in many ways, inimical or hostile to the environment. In a free market economy, competition breeds innovation, which, over time, reduces product prices and increases product quality. Consumers benefit from lower prices and higher quality goods; however, companies face continual downward price pressures, which reduce their overall gross revenue and profitability. **Equity valuations** (i.e., stock prices) of publicly traded companies are based on a perpetual growth model. Growth is fueled by selling the same products to new customers or more products to the same customers, thereby increasing market share. Growth is also a function of innovation or the creation and selling of new products to the same customer or new customers, thereby creating new markets. Companies that are able to demonstrate regular growth or the potential for future growth are rewarded with increasing equity valuations: their stock prices trend higher. Global capitalism thus rewards those companies willing to innovate, as innovation makes companies more efficient by reducing input costs and increases market share. This process, however, often results in increasing resource consumption, waste, pollution, and climate change. It is no surprise that, on a per-capita basis, the United States is the world's largest consumer of goods and services and also its biggest waste generator and polluter. The basic tension between globalization, market-based economies, and sustainability remains as economic growth in the current model cannot occur without significant **environmental costs**: The forces that increase human material well-being also destroy the environment.

Tesla, the world's most valuable automobile manufacturer in 2021, provides an excellent example of these inherent tensions. The **market capitalization** of Tesla in 2021 (stock price x outstanding shares) was more than the other three large auto manufacturers combined (29). This valuation was not based on the existing market share of Tesla automobiles, which was significantly less than any of the other traditional large automakers (e.g., GM, Ford), but its anticipated future market share. Tesla, though an EV (electric vehicle), poses a conundrum. Tesla automobiles don't emit carbon dioxide, a **greenhouse gas**; however, battery production and disposal have significant environmental costs. Additionally, Tesla batteries are often recharged using public utility electricity which, in the United States, is often derived from **nonrenewable sources** such as natural gas and coal. These are issues that can and will likely be addressed in the near term; however, they are illustrative of the dangers in relying only on technology to address environmental impacts. If we are to make any significant progress in combatting climate change, reducing pollution, and creating a healthy planet for not only humans but also other plant and animal life, we have to **reduce** consumption, as well as **reuse, recycle, or repurpose** existing products to divert them from the waste stream (30).

Possible Futures

No one knows what is going to happen with globalization as there is no one type of globalization happening. Rather, there are general trends (e.g., markets) that play out in specific ways (highly regulated markets or state directed markets) in different regions. In the early 1990s, Francis Fukuyama coined the phrase "the end of history" to suggest that humanity had finally arrived at the doorstep to a better life (31). The framework for this new age of man would be provided by enlightenment values (reason), democracy, free markets, technology, and a modicum of respect for the planet. The increasing amenities of urban life, the growth and penetration of technology in all aspects of our lives, and the spread of enlightenment ideals validate this view. The current parochialism, xenophobia, environmental problems, and conflict are seen by many as bumps on the road to a better future.

This **Panglossian** view, however, is simply not supported by history. Though we are approaching the 100-year anniversary of the last world war, the threats to global peace and stability, including climate change, are numerous. Humankind's ability to destroy itself and the planet is unprecedented. The weapons are multifaceted and include nuclear, biological, chemical, as well as virtual agents. The proliferation of nonstate actors with access to these weapons and their ability to coordinate, thanks to the tools of the internet, is also unprecedented. The almost overnight regression of Afghanistan to its pre-2000 Taliban roots after the end of the American occupation in 2021 and President Putin's veiled threats about using nuclear weapons in response to Western aid given to Ukraine should provide some pause to those who continue to proclaim the "end of history." Additionally, the Israel-Hamas conflict has the potential to grow regionally or beyond depending on how the various actors directly and indirectly involved react in the near future. Similarly, the 2008 recession, its aftermath, the mainstreaming of right-wing ideologies, the 2020 pandemic, and the global response all indicate a world that is **closer to midnight**, or its end, than we would care to acknowledge. Against this backdrop, climate change continues its inexorable march. Though climate conferences are held, new goals agreed on, and agreements announced, the modern, globalized, capitalist economy requires more and more of the planet. Though renewable energy sources have made significant inroads during the last few decades, much of the world will continued to be powered by fossil fuels—coal, petroleum, and natural gas—in the foreseeable future. Those pulled out of extreme poverty, the newly minted middle class, and the rapidly growing upper middle classes all aspire to the same creature comforts, the amenities, and the consumer products that those of us in developed countries now call rights. Our consumption and waste are likely to continue unabated. Yet, many different futures are possible; however, if we are to navigate these currents, all facets of society, from its formal institutions to informal groups, must be vigilant, collaborative, and proactive. The sacrifices must be borne by all, but also be paid for by those who polluted first and polluted most (i.e., those who industrialized first). The 2015 SDGs offer a template for government and state actors, but much can also be accomplished by the private sector and individuals.

Chapter 2

The Environmental and Health Impacts of Civilization

Climate Change, Resource Consumption, Waste

Key Terms

Anthropogenic: Environmental impacts related to human activity

Greenhouse gases (GHGs) and global warming: The gases in the atmosphere (carbon dioxide, methane, nitrous oxide) that trap heat and cause global warming, an increase in the planet's average temperature

Climate change: The various environmental impacts, such as extreme weather, associated with global warming

Energy: Essential for work in all aspects of civilization. Produced by both easily replenished renewable sources, such as wind or solar that have fewer environmental impacts, and nonrenewable sources, such as fossil fuels (coal, petroleum, natural gas) with significantly larger environmental consequences

Water: An essential ingredient for various life processes in most organisms. Water is also necessary for food production, electricity generation, sanitation, and various industrial functions.

WASH: An acronym for the relationship between water, sanitation, and hygiene.

water scarcity and stress: Inadequate water for various human needs resulting in economic, social and health related impacts. Scarcity and stress are the result of increasing population and changing consumption patterns (e.g., more meat eating or fast fashion leads to greater water use), environmental factors (pollution, climate change), and underdeveloped water infrastructure.

Malnutrition: Human health impacts related to insufficient food intake, including macronutrient (protein-energy malnutrition) and micronutrient (vitamins, minerals) deficiencies

Obesity: High BMI, or Body Mass Index (Weight in kilograms divided by height in meters squared), due to excess consumption of foods, especially processed foods or foods that have been produced or refined artificially

Waste: Discarded materials from human activity. Recycling can help divert items from the waste stream, reducing environmental impacts, and reclaim a portion of the original economic value. Waste includes nonmunicipal waste due to mining, agriculture, and industry, which is often handled *in situ* (where it was produced) and municipal solid waste (MSW), waste produced by various societal sectors (residential, commercial, industrial) and taken to landfills, openly dumped, incinerated, or recycled.

Other key terms will be in **bold** and defined in the text.

Review Questions

- What are the various greenhouse gases and their global warming potential, and how do they contribute to global warming?
- What are the sources of the different greenhouse gases?
- What are the impacts of climate change?
- What are the different sources of nonrenewable and renewable energy?
- What are the end uses (e.g., commercial, residential) for the various energy sources (e.g., nuclear, solar)?
- What are the various sources of fresh water?
- What is the relationship between water, energy, food, sanitation, and public health?
- What are the trends in food production? What types of food are being produced?
- What are the climate change impacts associated with food production?
- What are the public health impacts of food scarcity or excess food consumption?
- What is the relationship between population increase, economic growth, and waste production?
- What types of waste are recycled?
- What are the public health impacts associated with waste?

Introduction

As civilizations grow technologically and become more complex, their environmental footprint gets larger. The industrial revolution was an inflection point for humanity. Manufacturing increased the material standard of living and the size of the world's economy and unshackled many from agricultural labor. Resource consumption, waste, and pollution also increased to unprecedented levels. One major environmental impact was an increase in **greenhouse gases** (GHGs), such as carbon dioxide, nitrous oxide, and methane, that trap heat, increasing the earth's average long-term temperature, a phenomenon known as **global warming**. The various impacts, such as extreme weather, of global warming are collectively called climate change, pose an existential threat to many organisms, and will impact human populations in a multitude of ways, most of them negative.

Some climate change naysayers are skeptical that global warming is occurring. Some even argue that warming is beneficial for life on earth. Others concede that warming is occurring but deny it is **anthropogenic**, or due to human activity, and instead believe it to be part of a naturally occurring cycle. The scientific consensus, however, is that climate change is occurring and largely driven by human activity. Furthermore, the impacts of climate change are severe, potentially catastrophic, and impact every ecological niche. No creature, including man, is exempt from its deleterious impacts.

Climate change is not the only environmental issue linked to human activity. Energy and resource consumption, environmental pollution, and waste have increased markedly since industrialization. The cost of human civilization has, unfortunately, often been borne by the other nonhuman inhabitants of this planet. Many have succumbed to the environmental changes induced by humans and become extinct; many more are threatened and will soon cease to exist. Economically developed countries are the biggest resource consumers and polluters, though many have made important strides in the last few decades in reducing their environmental footprint. As other countries industrialize and develop economically, their environmental footprint will mirror that of developed nations.

Energy, water, and food are inextricably linked as large amounts of water are used to produce both energy and food. Hydroelectric dams use the mechanical energy, or movement, of water in turbines to produce electricity; thermoelectric plants use coal to turn water into steam to drive electricity-producing turbines; nuclear power plants also heat water to create steam. Water is essential for both agriculture and processed foods. Water is used directly for cooling and indirectly for HVAC systems, which use electricity often provided by hydro-, thermo,- or nuclear electric plants. Water is also used for nonfood-producing plants such as the front yard grass found in many American single-family homes. Water and sanitation are two sides of the same coin, as poor water management compromises sanitation, leading to greater health and environmental burdens. Water scarcity thus has the potential to upend several sustainable development goals given its integral role in plant and animal biological systems, food and energy production, and sanitation. Water is life giving and, during disasters, life taking. Climate change has led to both prolonged droughts in areas with little rainfall and heavier monsoons in areas with heavy precipitation. The negative impacts are inequitably distributed. The results have been disastrous for the most vulnerable humans—the children, poor, elderly, and disabled—and other organisms, many facing the threat of extinction.

Food, like water, is essential for life. Food production also illustrates the complex relationship between natural resources, economic growth, population growth, public health, and environmental degradation. Increases in agricultural yields due to the Haber–Bosch process, which extracts nitrogen from the atmosphere, and innovations in genetic engineering, pesticides, and plant management, collectively called the green revolution, have helped human populations grow by several billion since the mid 1800s. The demands placed on the food industry by the financial sector have led to new processed foods, many of which target children, larger portion sizes, and harmful additives for color, texture, taste, or preservation all in the name of increased sales. The public health impacts of more food and processed foods include an explosion of obesity and the attendant chronic illnesses such as heart disease, hypertension, and diabetes; evidence also links modern diets along with environmental pollution to several types of cancers. Modern food production systems have, ironically, liberated people from being tied to the land, granting them freedoms to explore other interests without fear of starvation and condemned them to food markets that are, in many ways, the largest contributor to their ill health. The environment has also suffered as food production has increased. Livestock bred for food and milk are the biggest source of methane, a GHG; food waste also generates a significant amount of methane; agricultural run-off increases both nitrogen and phosphorous in water streams, contributing to harmful algae blooms that deplete oxygen, decimating other water-based life.

This chapter provides an overview of global warming, a complex phenomenon that involves the interaction of solar radiation and GHGs in the atmosphere, which causes the many different impacts collectively called climate change. The chapter explores the impact of urbanization, population increase, and economic growth on increasing the utilization of key resources such as energy, water, and food and excessive waste production. The chapter also discusses disparities in impacts such as the greater exposure of the poor in developing regions to waterborne diseases, greater processed food consumption among the American poor, or greater susceptibility of the poor and minority populations to climate change–induced extreme weather in both developing countries and the United States.

Climate Change Overview

Climate and weather are sometimes used interchangeably though they are different. **Weather** refers to daily atmospheric conditions such as temperature, humidity, rainfall, or wind. **Climate** is the average weather over a longer span, typically 30 years or more. Similarly, global warming and climate change are colloquially used to refer to the same phenomenon, though climate change is one of the many impacts of global warming.

Climate change refers to the alterations in naturally occurring phenomena due to excess GHGs released into the earth's atmosphere as a result human activity. Climate change includes changes in ambient air and ocean temperature, sea level, acidity, precipitation, and extreme weather events (2). Increasing temperatures are also causing glaciers and sea ice to melt; glaciers are some of the largest reservoirs of fresh water on the planet. Ecosystem impacts are varied and include changes in plants' growing cycles and animal migration patterns.

Anthropogenic, or human-induced, climate change is different from **climate variability**, the short-term, naturally occurring variations in weather such as more or less snowfall in two consecutive years. Volcanic eruptions or asteroid collisions can cause climate change, as can unknown factors that trigger oxygen loss in oceans. The earth's history includes several periods of **naturally occurring climate change**, the most famous of which is the dinosaur extinction 66 million years ago likely caused by an asteroid. The last 500 million years have likely had five major **extinction events**, the largest of which occurred 252 million years ago, during the Permian era, and consumed nearly all known life on earth (3, 4). This extinction, known as the "great dying," was possibly triggered by volcanic eruptions in Siberia, significantly increased carbon dioxide concentrations in the atmosphere and oceans, and depleted dissolved oxygen in a relatively compressed time frame (4). Within 20,000 years, a heartbeat in geological time, 90% of land and ocean animals had perished as they were increasingly unable to breathe. Though CO_2 levels escalated rapidly, the rate of atmospheric accumulation in the Permian era was still below the rate due to human activity today (3).

Sunlight is essential for all life as it allows for photosynthesis, the process by which plants, photosynthetic bacteria, and a few other organisms use carbon dioxide and water to create sugar and oxygen ($6CO_2 + CH_2O = C_6H_{12}O_6 + 6O_2$). This process enables life on earth as nonphotosynthetic organisms consume plants and other photosynthetic organisms for energy. Sunlight is also the primary driver of the earth's climate. Sunlight, or solar radiation, contains both visible and ultraviolet light, half of which is either absorbed (20%) by the atmosphere or reflected by the earth's surface (30%). The remaining solar radiation (50%) is absorbed by the earth's surface and reemitted back into the atmosphere as infrared radiation. Most of the infrared radiation is absorbed by GHG in the atmosphere and reemitted in all directions. This process helps create a warmer climate by about 33 degrees Celsius than would occur in the absence of an atmosphere and greenhouse gases. The warmer temperatures due to GHGs have historically been life sustaining; however, the increases in GHGs due to human activity have led to excessive global warming as a larger proportion of solar radiation is being trapped within the earth's atmosphere. The **flora (plants) and fauna (animals)** we see today have evolved in a relatively narrow temperature range, which is being disrupted by climate change. Consequently, many are under stress, and some are at risk of extinction.

The earth's atmosphere is mostly (99.5%) comprised of gases that do not contribute significantly to global warming. A miniscule proportion of the atmosphere (0.43%) contains gases and water vapor with the ability to absorb infrared radiation reflected from the earth's surface. Gases implicated in climate change have varying **global warming potentials (GWP)**, or the capacity to warm the earth. Carbon dioxide has a GWP of 1 as it is the most abundant GHG and is thus used as the baseline gas for evaluating the GWP of other GHGs. The GWP of other gases indicates how much radiation 1 ton of the gas will absorb relative to 1 ton of carbon dioxide in 100 years. A higher GWP indicates a greater warming potential relative to carbon dioxide.

Earth's average ocean and land temperature has increased by about 2 degrees F since the start of the industrial revolution in the mid-1800s (5). Carbon dioxide concentrations have increased by nearly half, from 280 parts per million by volume (PPMV) since before 1750, or pre-industrial times, to 416 ppmv today (2020). Similarly, methane concentrations have increased by one and half times since pre-industrial times, from 270 parts per billion (ppb) to 1879 ppb; and nitrous oxide concentrations have gone up by nearly a quarter (23%), from 222 ppb during pre-industrial times to 270 ppb today (5). If current levels of GHG emissions continue, the earth's average temperature is likely to increase by up to 8 degrees F; even the most optimistic scenarios foresee a rise of at least 0.5 degrees F.

Ozone is another important gas in the atmosphere that is produced naturally and impacted by human activity. Ozone is composed of three oxygens (O_3) and forms when ultraviolet radiation splits atmospheric oxygen (O_2), creating a free oxygen molecule that binds with O_2 and creates ozone, or O_3. The newly created ozone molecule absorbs UV energy and splits into O_2 and a free O, which joins another O_2, restarting the cycle (59). This process helps shield the earth from harmful UV radiation; however, ozone in the lower atmosphere can negatively impact the human respiratory system, causing airway and lung inflammation and, in some cases, increase mortality risk, especially among the elderly. Chlorofluorocarbons (CFC) produced by human activity were reducing ozone concentration in the atmosphere, creating the infamous "holes in the ozone layer." Given the role ozone plays in absorbing harmful UV radiation, CFCs became heavily regulated in the 1990s, leading to a decrease in their atmospheric concentrations (59). Today, ozone is on track to recover completely by midcentury thanks to the strict regulation of CFCs and other harmful ozone-depleting chemicals.

A breakdown of the earth's atmosphere is provided in Table 2.1, along with the GWP of the various gases, common sources, and approximate annual discharge.

Gas	Percentage	GWP	Sources
Nitrogen	78.1%	—	
Oxygen	20.9%	—	
Argon	9%	—	
Carbon dioxide (CO$_2$)	0.39%	1	Fossil fuels combustion, forest clearing, biomass burning, production processes (e.g., cement)
Methane (CH$_4$)	0.00018%	25-30	Landfills, agriculture (livestock, rice cultivation), coal mining, oil and natural gas production
Nitrous oxide (N$_2$O)	0.000032%	270-300	Agriculture (e.g., fertilizer, nitrogen-fixing crops), livestock waste, biomass combustion, waste incineration
Sulfur hexafluoride	6.7 x 10 -10%	22,800	Human made, industrial and manufacturing processes
Chlorofluorocarbons (CFCs), Hydrofluorocarbons (HFCs) Hydrocholorfluorocarbons (HCFCs) Perfluorocarbons (PFCs)		Up to 14,800	Human made, industrial and manufacturing processes

Adapted from Environmental Protection Agency (EPA), "Understanding Global Warming Potentials," April 2, 2023, https://www.epa.gov/ghgemissions/understanding-global-warming-potentials

Greenhouse Gases

Globally, carbon dioxide from fossil fuel combustion and industrial processes accounts for the majority of emissions (65%), followed by carbon dioxide from forestry and other land uses (11%); methane (16%) from agriculture and waste management activities; nitrous oxide (6%) from agriculture; and other gases (2%) from industrial processes, refrigeration, and the use of various consumer products (18). Electricity and heat generation (25%), along with agriculture, forestry, and other land uses (24%) account for nearly half of global GHG emissions. Industries (21%) are the next biggest emitters, followed by transportation (14%), other energy-related activities (10%), and buildings (6%) (18, 19). The largest emitters are China (30%), the United States (15%), the European Union (9%), India (7%), the Russian Federation (5%), and Japan (4%) (20). On a per-capita basis, the United States has the largest GHG footprint. It emits 14 tons of CO$_2$ per capita annually. In comparison, China emits 8.4 tons, which is still more than the 8.2-ton per capita average for advanced economies, and 40% higher than the 6-ton EU average (21). The United States is also the world's largest energy consumer and overall waste producer on a per capita basis.

COVID-related shutdowns helped drive down overall GHG emissions in the United States by 9% in 2020. As shown in Table 2.2, total GHG emissions in 2019 were roughly the same as in 1990 and 12% lower than 2005 (5). Carbon dioxide (-7.9%), methane (-16.7%), and nitrous oxide (-5.4%) emissions decreased between 1990 and 2020;

however, fluorinated gases increased by nearly 90%. These gases, though emitted in extremely small amounts relative to the three big gases (CO_2, CH_4, N_2O), have much higher global warming potentials. As shown in Table 2.2, the increase in GHG concentrations from pre-industrial (1750) times to today indicates that anthropogenic, or human activity, plays a decisive role in global warming observed today.

Table 2.2: GHG Emissions by Gas in MMT CO_2 Eq. and Concentration Increase of the Three Most Common GHGs Since Pre-Industrial Times

Gas	1990	2005	2016	2019	2020	2020 % of total GHG	% Conc. 1750 to 2020
CO_2	5,122	6,137	5,252	5,259	4,715	78.8%	49% +
CH_4	781	698	658	669	650	10.9%	157% +
N_2O	451	453	449	457	426	7.1%	23% +
Other GHGS	100	146	179	85	189	3.2%	
Total	6,454	7,435	6,538	6,572	5,981		
Percentage change		15.2%	-12.1%	0.5%	-9.0%		

Adapted from EPA, "Inventory of U.S. Greenhouse Gas Emissions and Sinks: 1990–2020," 2022, https://www.epa.gov/ghgemissions/draft-inventory-us-greenhouse-gas-emissionsand-sinks-1990-2020; IPCC, "Climate Change 2021: The Physical Science Basis." Working Group I to the Sixth Assessment Report of the Intergovernmental Panel on Climate Change, 2021, https://www.ipcc.ch/report/ar6/wg1/; NOAA/ESRL, "Trends in Atmospheric Carbon Dioxide, Methane, Nitrous Oxide," https://gml.noaa.gov/ccgg/trends/gr.html.

Table 2.3: GHG Emissions in Kilotons (kT) in 1990, 2005, 2019, and 2020

GHG	1990	2005	2019	2020
CO_2	5,122,496	6,137,603	5,259,144	4,715,691
CH_4	31,233	27,898	26,753	26,017
N_2O	1512	1,521	1,533	1,430

Adapted from EPA, "Inventory of U.S. Greenhouse Gas Emissions and Sinks: 1990–2020," 2022, https://www.epa.gov/ghgemissions/draft-inventory-us-greenhouse-gas-emissionsand-sinks-1990-2020; IPCC, "Climate Change 2021: The Physical Science Basis." Working Group I to the Sixth Assessment Report of the Intergovernmental Panel on Climate Change, 2021, https://www.ipcc.ch/report/ar6/wg1/; NOAA/ESRL, "Trends in Atmospheric Carbon Dioxide, Methane, Nitrous Oxide," https://gml.noaa.gov/ccgg/trends/gr.html.

CO_2 emissions accounted for the bulk (78.8%) of GHG emissions in the United States in 2020. Methane (CH_4) emissions were a tenth (10.9%) of total GHG emissions and came primarily from domestic livestock for which methane was a byproduct of enteric fermentation processes; waste decomposition and natural gas systems were also important sources of methane (5). Agriculture and wastewater treatment were major sources of nitrous oxide, which was 7.1% of total GHGs emitted in 2020 (5). The electronics industry, heating and cooling systems, electrical transmission infrastructure, and primary aluminum production were responsible for the other gases (e.g., CFC, SF6, NF3) with high GWP. These high GWP gases collectively accounted for 3.2% of the total GHGs emitted in the United States in 2020. Water vapor also plays an important role in naturally occurring greenhouse processes. Unlike other GHGs, high concentrations of water vapor lead to precipitation, which decreases concentration, creating an equilibrium. Though some human activities such as agricultural watering or the cooling of power plants increase water vapor concentration, these impacts have a negligible impact on climate change.

Table 2.4 provides a snapshot of the latest U.S. GHG emission data in CO_2 **million metric ton (MMT)** equivalents using IPCC categories (a million metric tons is roughly 2.2 billion pounds). The **IPCC, or Intergovernmental Panel on Climate Change,** was created in 1988 by the World Meteorological Organization and the United Nations to provide analysis, recommendations, and guidance to governments on climate-related policies. As part of its work, the IPCC evaluates current research on climate change, including its mechanisms (i.e., the science of climate change), likely impacts (environmental, social), and strategies to address its various impacts. To date, the IPCC has produced five assessment reports, the most recent in 2022.

The combustion of fossil fuels for energy was the largest source of GHGs in the United States. Within the combustion category, the two largest sources of GHGs were transportation and electricity generation. Other major sources of GHGs in the energy category were natural gas systems and coal mining. The largest sources of GHGs in the industrial processes category were iron and steel production, followed by cement production. Soil management and enteric fermentation (gases released by livestock) were the biggest GHG emitters in the agricultural category; landfills and wastewater treatment were the biggest sources in the waste category.

Table 2.4. United States GHG Emissions by IPCC Categories (MMT CO_2 Eq.)

Sector	1990	2005	2019	2020
Energy	5,341	6,320	5,410	**4,855 (81%)**
Fossil fuel combustion	4,731	5,752	4,852	4,342
Transport	1,469	1,859	1,814	1,572
Elec. Gen.	1,820	2,400	1,606	1,439
Industrial	854	852	816	766
Residential	339	359	341	316
Commercial	228	227	251	227
U.S. Territories	22	56	24	23
Natural gas	227	203	211	200
Coal mining	101	68	50	43
Industrial processes	346	366	380	**376 (6%)**
Iron/steel production	105	70.1	43	38
Cement	34	46	41	41

Agriculture	552	574	623	**595 (10%)**
Soil management	316	314	345	316
Enteric Fermentation	164	168	176	175
Waste	214	176	160	**156 (3%)**
Landfills	177	132	114	109
Wastewater treatment	37	41	42	42
Total	6,463	7,436	6,573	**5,982 MMT CO2**

Adapted from EPA, "Inventory of U.S. Greenhouse Gas Emissions and Sinks: 1990–2020," 2022, https://www.epa.gov/ghgemissions/draft-inventory-us-greenhouse-gas-emissionsand-sinks-1990-2020.

Climate Change Impacts

Risk, or the potential for adverse consequences, to human and other ecologies is the basis for assessing climate changes impacts by the IPCC (17). Risk is a function of **hazards**, or the potential for the occurrence of an event with deleterious consequences for humans and other ecologies, **exposure**, or the presence of vulnerable human and nonhuman populations in impacted areas, and **vulnerability**, or the higher likelihood of being impacted negatively because of biological, social, and other reasons (17). **Adaptation and resilience** are other factors assessed by the IPCC, as both play a role in reducing risk or responding in ways that reduce adverse impacts.

Climate change has numerous impacts across different ecosystems. Some of the impacts are direct, and others are mediated by factors exacerbated by climate change. In some cases, the impacts cause irreversible damage to ecosystems and human communities. The most vulnerable among humans are often those from marginalized communities such as those living in extreme poverty in developing countries or, in the United States, lower income groups and some racial and ethnic minorities. Many of the observed impacts are noted in Table 2.5. In the absence of significant ameliorative steps taken by all countries, especially the biggest emitters such as China and United States, the risks associated with climate change will compound, or magnify each other, as exposure and vulnerability will both increase, as will a reduced capacity for adaptation and resilience. Technology provides humans with an advantage not available to other nonhuman inhabitants of the planet, many of whom will fare badly and, possibly, perish. Though humans may survive, they will not thrive, given the sometimes pyrrhic costs associated with relying on technology to address issues that could have been prevented.

Some of the risks identified by IPCC are noted in Table 2.5. This is not an exhaustive list, as many risks are still not well understood, and others will emerge if climate change continues unabated. Additionally, risks intersect with each other in complex ways, sometimes amplifying each other, depending on context. For example, climate change increases the risk for hot extremes on land and drought. In Southern California this increases the risk for wildfires and water scarcity. In developing countries, extreme heat increases malnutrition risk among the poor as it reduces agricultural productivity. Additionally, the poor have a higher risk for heat-related health conditions as they often inhabit informal dwellings without proper insulation or mechanisms for artificial cooling (e.g., air conditioning) and work in professions with greater exposure to the environment. Urban encroachment also reduces biodiversity by claiming land for human habitation and by increasing resource consumption, waste, GHGs, and related risks such as wildfires or runoff, which independently impact biodiversity.

Observed Impacts	Degree of Confidence VHC = Very high confidence, HC = High, MC = Medium
Increasing frequency and intensity of climate and weather extremes, including hot extremes on land and in the ocean, heavy rain, drought and fire weather	HC
Heat-related public health impacts	MC
Heat-related tree mortality	HC
Negative impacts associated with tropical cyclones	MC
Ocean acidification, sea level rise	HC
Irreversible damage to terrestrial, freshwater, coastal, and open-ocean marine ecosystems	HC
Shifts in seasonal timing	HC
Mass mortality events on land and oceans	VHC
Hydrological changes due to glaciers melting and changes in mountain and arctic ecosystems due to permafrost thaw	HC
Reduced food and water security for human populations; increasing malnutrition	HC
Reductions in agricultural productivity	MC
Direct impacts on physical health and mental health, including increases in vector-borne illness, waterborne diseases, zoonotic diseases, and heat-related illnesses	VHC
Adverse impacts on key infrastructure (water, waste, transportation, etc.)	HC
Impacts are disproportionately felt by marginalized populations	HC
Humanitarian crises are worsened, increasing displacement of human populations	HC

Adapted from IPCC, Summary for Policymakers. In Climate Change 2022: Impacts, Adaptation and Vulnerability, 2022, Cambridge University Press. https://doi.org/10.1017/9781009325844.001.

Water, Sanitation, and Hygiene

Sustainable Development Goal (SDG) 6 is focused on "ensuring the availability and sustainable management of water and sanitation for all." Safely managed water service refers to "improved drinking water facilities on premises (e.g., water tap), which provides continuous, uncontaminated, drinking water." Safely managed sanitation includes latrines that are "not shared with other households and excreta that are either treated and disposed of *in situ*, stored temporarily (e.g., septic tank) and taken for treatment off-site, or transported through a sewer system, along with wastewater, for treatment off-site (26). Hygiene includes cessation of open defecation when alternatives are available and practicing regular handwashing with soap when possible. Good hygiene practices are instrumental in preventing respiratory infections among children and a host of other infectious diseases.

The UN estimates that around 3 billion people rely on a water source whose quality is not regularly monitored and over 700 million live in regions with high or critical levels of water stress. In 2020, 2 billion people did not have access to safely managed drinking water and nearly twice that, or 3.8 billion, lacked safely managed sanitation. If current trends continue, by 2030, 1.6 billion will still lack access to safe drinking water, and 2.8 billion will lack adequate sanitation (34). Poor water management exacts significant public health and environmental costs across the planet as waterborne disease is a major source of preventable **morbidity** (sickness) and **mortality** (death) in the world, particularly in developing regions. The World Health Organization (WHO) estimates that 3.3% of global deaths annually, or around 2 million people, and 4.6% of DALYS (disability-adjusted life years), or 123 million DALYs, are directly attributable to WASH-related issues (26). The burden among children under 5 is severe, as 13% of deaths and 12% of DALYS could be prevented through provision of adequate water and sanitation services and basic hygiene practices (26). Poor water treatment or sanitation is a major factor in the transmission of preventable diarrheal diseases, which are the second leading cause of death for children under 5, with 8% of deaths (over 500,000) annually attributed to diarrhea-causing diseases such as cholera, dysentery, typhoid, and tape worms; 700 children under 5 perish daily from preventable water- and sanitation-related diseases; nearly a half billion children live in areas with severe water shortages (26,2 7). Diarrheal diseases can be caused by bacteria (e.g., shigella), viruses (e.g., rotavirus), or parasites (e.g., ringworms) and are typically spread through ingested fecal matter.

In 2020, nearly 80% (6.1 billion) of people had some access to basic sanitation, and just over half (54%, or 4.2 billion) had access to a safe sanitation service (28). However, 1.7 billion (22%) did not have access to any basic sanitation such as a private toilet (28). It is estimated that between 6% and 11% of the world's population (500–900 million) still practices open defecation, which increases the risk for fecal matter–related diseases, including soil-transmitted helminthiasis or worms (29). Worldwide, nearly half (45%) of household wastewater is discharged without treatment, and a tenth of the world's population consumes agriculture irrigated by wastewater (36). Several different worm types (ascariasis or roundworm, trichiniasis or whipworm, ancylostomiasis or hook worm) are transmitted from soil contaminated with feces; direct feces or human to human contact is not implicated in helminthiasis (29). Severe infections can significantly impair the physical and cognitive development of infected children, many of whom suffer from repeat infections. For women, as they are often alone when engaging in the practice, open defecation can increase their risk for assault and other predatory behaviors. Malaria is the single biggest vector-borne disease linked to poor water management as mosquito larvae thrive in clean, unpolluted, stagnant water. Annually, over 200 million people suffer from malaria, and nearly half a million succumb to it, most of whom live in sub-Saharan Africa. There are many other illnesses, such as schistosomiasis, dengue, and hepatitis A, whose prevalence can be drastically reduced through WASH-related strategies. Figure 2.1 shows how inadequate water infrastructure can ultimately impact economic development, which perpetuates poor water infrastructure. Children who are sicker (i.e., higher morbidity) will be stunted physically and mentally, compromising their future economic prospects. If the problem is pervasive, as it is in many developing regions, economic development at the societal level will be negatively impacted, creating a feedback loop that will perpetuate poor infrastructure.

Though waterborne illnesses are common in developing countries, wealthier countries are not immune from health burdens related to poor water management. For example, the United States had over 6,000 deaths in 2014 from waterborne pathogens. The morbidity burden was also high, with over 7 million illnesses, 600,000 ER visits, and 118,000 hospitalizations attributed to waterborne illnesses (30). The vast majority of water-related illnesses and ER visits were related to swimmers' ear or otitis externa; however, hospitalizations and deaths were mostly due to nontuberculous mycobacterial infections (30). The 2014 Flint water crisis, in which drinking water was contaminated with lead and bacteria, and the more recent 2022 Jackson water crisis, which required residents to go without running water on some days and boil water for a month, highlight the importance of well-maintained water infrastructure. Both water-related emergencies also occurred in predominantly Black communities, illustrating the disproportionate burdens often borne by minority and other marginalized communities in the United States. A report by the Michigan Civil Rights Commission argued that implicit bias and failure to evaluate the disproportionate impacts of policy on minority communities (i.e., systemic racism) by policymakers contributed significantly to the Flint water crisis. Similarly, some have argued that the Jackson water crisis is the result of systemic racism and includes infrastructure funding decisions at the state level that privilege White communities in Mississippi at the expense of majority Black communities such as Jackson (31, 32, 33).

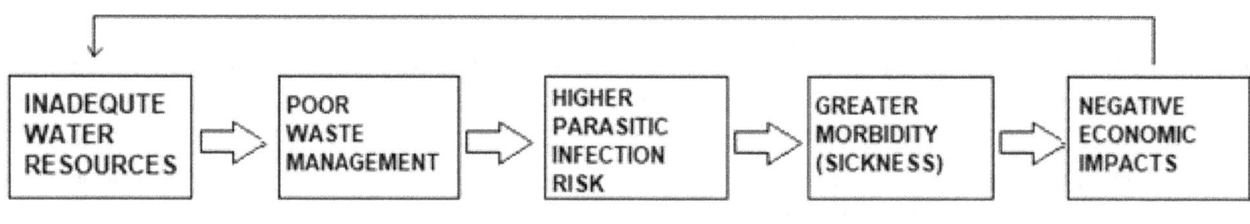

Figure 2.1　The relationship between water, sanitation, and economic development.

Water Use and Availability

Water's unique molecular structure, consisting of two hydrogen atoms and one oxygen (H_2O) atom, makes it a vital part of biological systems. The biggest component of many organisms, including humans, is water. Water is essential for photosynthesis, which allows plants to produce sugars that power all of life on earth. The uneven distribution of electrons in a water molecule gives its oxygen atom a negative charge and hydrogen atoms a slight positive charge. These dipoles, or different charges, give rise to hydrogen bonding, in which the positively charged hydrogen is weakly bonded to a negatively charged oxygen in another water molecule. (Covalent bonds between H and O in a single water molecule are much stronger; see Figure 2.2.) Hydrogen bonding plays an important role in **capillary action**, the movement of water molecules up a plant against gravity, water's ability to act as a nearly universal solvent, and its lower density as a solid. Water's capacity for dissolving other compounds makes it necessary for many life processes within organisms. Unlike other compounds, water is less dense as a solid, allowing bodies of water to freeze top down. Even during the frigid winter months, life thrives below the surface, where water is still in its liquid state. Water is also an important structural element within cells and in cell membranes.

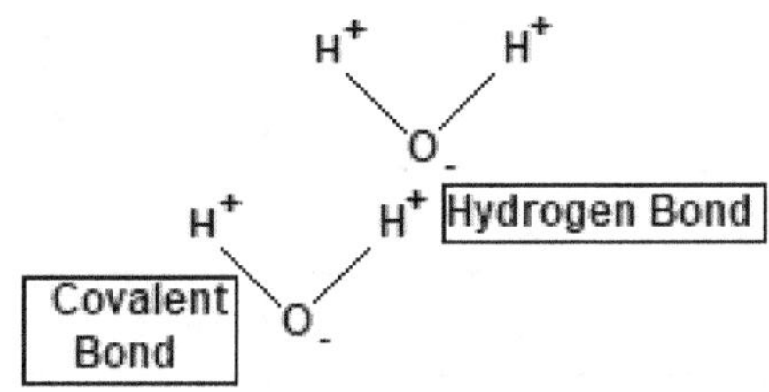

Figure 2.2   Hydrogen and covalent bonding in water.

Worldwide, agriculture, which includes animal husbandry, consumes the most water. Tables 2.6a–b and 2.7 provide an overview of the water needed for common foods, beverages, and crops. As shown in the tables, food production is very water intensive. Fruits and vegetables have the lowest water needs, and animal sources foods (ASF), such as beef and lamb, along with processed foods, such as chocolate and butter, require the largest amounts of water. The crops with the highest caloric efficiency, or the most energy provided for every unit of water, are sugars, cereals, and oils; those with the lowest caloric efficiency are nuts, fruits, and spices. As countries move up the income ladder, consumption of animal source foods and processed foods tends to increase, as does energy use, all of which increase water consumption substantially. Ruminant meat demand is expected to almost double (88%) between 2010 and 2050 (42). This shift will not only increase water consumption but also GHGs as ruminants produce methane.

Food	Liters per kg	Liters per lb	Liters per oz	Gallons per oz
Chicken	4,325	1,966	123	32
Beef	15,400	7,000	438	116
Sheep	10,412	4,733	296	78
Pork	5,988	2,722	170	45
Rice	2,417	1,099	69	18
Lettuce	240	109	7	2
Tomato	214	97	6	2
Potato	287	130	8	2
Maize	1,222	555	35	9
Peach	910	414	26	7
Orange	560	255	16	4
Apple	822	374	23	6
Banana	790	359	22	6
Mango	1,800	818	51	14
Olives	2,015	916	57	15
Peanuts	2,782	1,265	79	21
Pasta	1,849	840	53	14
One egg	196	89	6	1
Chocolate	17,196	7,816	489	129
Cheese	3,178	1,445	90	24
Butter	5,553	2,524	158	42
Wheat Bread	1,608	731	46	12

Adapted from Water Footprint Network, "Product Gallery," https://waterfootprint.org/en/resources/interactive-tools/product-gallery/.

Beverage	Liters per 250 ml(8.5oz)	Gal per 250 ml (8.5oz)
Beer	74	20
Wine	218	58
Tea	27	7
Coffee	264	70
Milk	255	67

Adapted from Water Footprint Network, "Product Gallery," https://waterfootprint.org/en/resources/interactive-tools/product-gallery/.

Table 2.7. Water Need by Crop Category (Total and Liter per kCal) and Caloric Efficiency (kCal per kg)

Crop Category	Water (m3 per Ton)	Kcal/Kg	Liter/kcal
Sugar crops	197	290	0.68
Vegetables	322	240	1.34
Fruits	967	460	2.10
Cereals	1,644	3,200	0.51
Oil crops	2,364	2,900	0.81
Pulses	4,055	3,400	1.19
Spices	7,048	3,000	2.35
Nuts	9,063	2,500	3.63

Adapted from The Green, Blue and Grey Water Footprint of Crops and Derived Crop Products, by M.M. Mekonnen & A.Y. Hoekstra, Earth Syst. Sci, 15, 1577–1600.

Though agriculture is the largest water use category at the planetary level (69%), water use and availability vary greatly by country. Industry, which includes electricity generation, accounts for 19% of water use worldwide, with municipal use at 12% (47). Generally, per-capita water use is higher among developed nations, where water is also used in large quantities for nonfood-related purposes such as energy production (e.g., thermoelectric turbines), cooling, industrial processes, and recreation. For example, as shown in Table 2.8, the largest uses of water in the United States are thermoelectric power generation (41%), followed by agriculture (37%). In Europe, water used for cooling power plants accounts for 43% of freshwater withdrawals (37). Most (90%) power that is generated worldwide requires water for some part of the electricity generation process (37).

In 2015, the average per-capita water use in the United States was 82 gallons, or 310 liters (25). The average U.S. family spent around $1,000 per year on water in 2015, or about 2% of median family income (25). The UN recognizes safe and affordable water access as a fundamental human right: Every human being should have daily access to between 50 and 100 liters of water for personal and domestic use; water sources must be located within 1,000 meters of the home; collection time should be within 30 minutes; and water cost should not exceed 3% of household income. However, despite improvements in water infrastructure worldwide, 2 billion people in 2020 did not have access to safely managed drinking water.

Table 2.3. U.S. Daily Water Withdrawals (2015) in Billions of Gallons per Day

Category	Billions of Gallons per Day	Percentage of Total
Public supply	39	12%
Self-supplied domestic	3.26	1%
Irrigation	118	37%
Livestock	2	1%
Aquaculture	7.55	2%
Self-supplied Industrial	14.8	5%
Mining	4	1%
Thermoelectric	133	41%
Total	322 billion gal per day 87% fresh water	

Adapted from "Estimated Use of Water in the United States in 2015," by C. A. Dieter et al., U.S. Geological Survey Circular, 1441, 2018, https://doi.org/10.3133/cir1441.

Though water covers much of the world's surface area (71%), it is only a small portion of the total planet's volume or mass. The water cycle moves water from various surface sources, through evaporation or transpiration, to the atmosphere, where after 5 to 9 days it returns to the surface as rain or snow (precipitation). Water can stay as long as several thousand years in ice caps or ground water or as little as a few hours inside living organisms (43). Irrespective of where water is located, it will eventually change locations. Only 3% of the world's total water (332.5 million cubic miles) is fresh water, almost all of which (99%) is difficult to access as it's sequestered in either ice caps and glaciers (69%) or groundwater (30%). Surface water is the most visible freshwater source; however, it accounts for less than 1% of the planet's freshwater supply: Rivers (.0006%) and lakes (.26%) together account for less than a third of 1% of fresh water. Precipitation that renews freshwater sources is also unequally distributed with the highest rainfall rates occurring in Brazil, Russia, Canada, Indonesia, China, Colombia, the United States, Peru, and India (45).

Water-Related Issues

Water stress is the result of three factors: increasing populations in regions with lower water availability (precipitation and storage), rising per-capita demand (due to increasing energy use and animal source food consumption as incomes rise), and climate change–related effects. The greatest needs are in places with burgeoning populations and underdeveloped water storage and infrastructure, such as parts of Africa and Asia. As shown in Table 2.9, Asia has 35% of the world's liquid freshwater; however, it also has the lowest per-capita availability. Africa will experience the biggest population growth in the coming years, due to high fertility rates; however, despite having high freshwater availability, water infrastructure is underdeveloped. Precipitation and ground water storage capacity are sometimes at odds in Africa. For example, Western and Central African countries have high precipitation but low groundwater storage capacity; the opposite is true in Eastern and Southern Africa. The development of groundwater storage capacity would also significantly augment economic development in parts of sub-Sahara Africa (46).

Table 2.9. Total Water Availability and Per-Capita Water Availability by Region and Projected Population 2050

Region	Km^3 x 1,000 Water	% Total Water	2021 Pop	% 2021 Pop	Km^{3}x1000 per Million People	Projected 2050 Population	% 2050 Population
North America	2071	19%	597	8%	3.5	696	7%
South America	1299	12%	434.3	6%	3.0	491	5%
Europe	541	5%	747.8	9%	0.7	710	7%
Africa	2722	26%	1373.5	17%	2.0	2489	26%
Asia	3691	35%	4679	59%	0.8	5290	54%
Australia & Oceania	324	3%	43.2	1%	7.5	57	1%

Adapted from "World Water Resources by Country (Table 3)," Food and Agricultural Organization, https://www.fao.org/3/Y4473E/y4473e08.htm.

Climate change exacerbates the spatial mismatch between water need and availability by increasing the frequency and intensity of droughts in dry areas and increasing precipitation in wet areas as warmer air can hold more water. Today, almost 2 billion people, most of them (73%) in Asian countries, live in water-scarce areas (44). Global water use is at an all-time high, nearly six times more than a century ago, and will increase by at least 20% by mid-century (44). Better water management strategies and steps to reduce climate change, along with the development of water infrastructure and resources, such as groundwater, are critical to reducing future water scarcity and preventing conflicts over water resources, especially **transboundary resources,** or water sources that cross national boundaries such as the Nile.

Other important issues related to water include the management of water pollution and water-related disasters. Climate change has increased the frequency and intensity of hurricanes and monsoons, most recently Hurricane Ian, which made landfall in the southeastern United States in September 2022. The estimated damage from the hurricane is likely to be in the hundreds of billions. In the United States, though insurance is available for postdisaster recovery, insurers sometimes cannot settle all claims given the extent of the damage, and the government must provide emergency aid thru FEMA. In many lower- and middle-income nations, both private (insurance) and public sector (government aid) postdisaster responses are severely inadequate. The brunt of the economic and other damage (e.g., health related) is often borne by the poorest members of society; this pattern exists in both developing and developed countries. Environmental-related inequalities are discussed more extensively in Chapter 8 in the environmental justice section.

Table 2.10 provides a summary of different water pollutants, the human activities that engender them, and their impacts. As shown in the table, the types of pollution are numerous and the impacts wide ranging. In many developing countries, there is little regulation and water treatment minimal or nonexistent with the poor, not surprisingly, bearing the heaviest health-related costs. Even in developed countries, like the United States, regulation is a relatively recent phenomenon dating to the passage of environmental legislation in 1948 (Water Pollution Act) and during the Nixon era in the 1970s (e.g., Clean Water Act, 1972).

Table 2.10. Water Pollutants, Sources, and Impacts

Pollutant	Source	Impact
Organic waste that is decomposed by microorganisms using oxygen	Sewage, industry runoff	Depletion of dissolved oxygen (DO) in water, which kills marine life
Agricultural runoff that includes inorganic compounds such as nitrogen and phosphorous	Sewage, agriculture, and urban runoff	Algae blooms and growth of other undesirable plants that deplete dissolved oxygen (DO) in water, killing other marine life
Thermal (heated) effluent water	Power plants and industries that use water for cooling and pump heated water back into the environment	Depletion of dissolved oxygen (DO) in water and harm to colder water organisms
Sediments and suspended particles such as silt, clay, debris, plankton	Urban runoff	Reduction in water clarity and sunlight entering water, which can impact bottom-dwelling life
Synthetic, volatile organic chemicals such as oil and pesticides	Agricultural runoff	Water taste is altered, freshwater plants can suffer
Inorganic chemicals such as acids and heavy metals	Industrial effluent, mining runoff, air pollution	Toxic to marine life and humans
Radioactive substances	Nuclear waste, medical waste, industrial waste	Toxic to marine life and humans
Bacterial contamination	Sewage (human waste), other bacterial sources	Increase in waterborne diseases, including diarrheal diseases

Adapted from Environmental Land Use Planning and Management by J. Randolph, 2012, Island Press.

Energy

Energy (e.g., heat, light, mechanical, chemical, etc.) is used to do **work**, the application of force over some distance. The law of conservation of energy asserts that energy can only change form and can neither be created nor destroyed. This is the basis for electricity generation or transportation as chemical energy in fossil fuels such as coal or petroleum is converted into mechanical energy used to power auto engines or a turbine generating electricity. Energy is essential for the sustenance and development of human civilization as it powers the various technologies that enable societal differentiation, freeing up much of humanity to engage in tasks unrelated to the provision of essentials such as water, food, or shelter. In pre-industrial societies, animals, slaves, and others in servitude provided much of the labor that today is tasked to machines, which have either replaced humans wholesale or made them more efficient. The improvements in AI will likely accelerate this trend further. The energy dependence of civilization was the basis of Nikolai Kardashev's scale for estimating a civilization's level of technological development. The Kardashev Scale classifies civilizations by their energy consumption: The more advanced a civilization, the higher its energy footprint. The initial scale used three levels, with the first level (Type I) corresponding to earth's energy usage in 1964. A Type II civilization could harness the energy of its star, and a Type III would harness energy at a galactic level (22). The scale has since been expanded, with humanity scoring a 0.7, or almost a Type I civilization. Some speculate that the transition to a Type I and beyond would carry a high risk of self-destruction as the home planet's (e.g., Earth) energy sources would be depleted, a process that would also wreak potentially catastrophic environmental damage. The energy profiles of the United States and most other countries today lend credence to this argument as most energy today comes from sources that add pollutants and GHGs to the atmosphere, warming the planet and contributing to environmental degradation in a multitude of ways.

Energy can be obtained from either nonrenewable or renewable sources. **Nonrenewable sources** are primarily **fossil fuels** such as **petroleum, coal, and natural gas** formed from long dead animals and plants. These sources of energy are limited in supply and usually need to be extracted or mined. **Nuclear energy** obtained through **fission**, or the splitting of atoms, is often classified as nonrenewable as the main fuel source is uranium, which also must be mined. **Renewable energy** sources include solar from the sun; geothermal, the heat from the earth; wind; biomass from plants; and hydropower from the movement of water. Table 2.11 provides a breakdown of U.S. electricity generation by renewable and nonrenewable sources. As shown in Table 2.11, 87% of the energy in the United States is obtained from nonrenewable sources (23).

A turbine converts thermal or mechanical energy supplied by a primary energy source such as coal or wind into electrical energy, which is then transmitted through the grid to consumers. A turbine works on the principle of **induction**, discovered by Michael Faraday in 1831, in which a moving magnet inside a wire coil induces electricity in the wires. All primary sources create rotation in the turbine by either using steam, gas, wind, or water. **Steam turbines** are the largest sources of electricity in the United States (44%) and the world (24). The largest use of water in the United States is also for electricity generation, as water is heated to produce steam. This method of electricity generation is different than hydroelectric processes, which use the mechanical movement of water in turbines to generate electricity. Air, like water, is a fluid whose movement can be similarly used to convert mechanical into electrical energy through a turbine. Solar energy provides electricity and is essential for photosynthesis, the process by which plants turn carbon dioxide and water into sugars, thus providing the first link in the food chain. Solar energy can be directly converted to electricity using photovoltaic cells, which are used in devices such as calculators or placed on rooftops to power buildings. Solar thermal power generators, in contrast, use solar energy to heat a fluid to generate steam for use in a turbine.

Table 2.11. U.S. and Worldwide Energy Sources (Total Energy and Electricity Generated), 2021

SOURCE	(%) of Total Energy (U.S.)	(%) of Total Energy (World)	(%) of Electricity Generated (U.S. only)	(%) of Electricity Generated (World)	Type
Petroleum	36%	30.9%	0.5%	2.8%	Non
Natural gas	32.2%	23.2%	38%	23.6%	Non
Coal	10.8%	26.8%	22%	36.7%	Non
Nuclear	8.4%	5%	19%	10.4%	Non
Biomass	5%	9.4%	1.3%		Renewable
Wind	3.4%		9.2%		Renewable
Hydropower	2.3%	2.5%	6.3%	15.7%	Renewable
Solar	1.5%		2.8%		Renewable
Geothermal	0.2%		0.4%		Renewable

Table 2.12 shows the relative contribution of different energy sources to each of major end use sectors: transportation, industrial, residential, commercial, and electric power generation. The first percentage indicates the proportion of the power source allocated to the end use sector; the second indicates the percentage of the end use sector that is powered by the energy source. For example, 69% of petroleum is used for transportation; however, 90% of the energy for transportation comes from petroleum. The electric power sector sells electricity and heat to consumers. Coal is mostly (90%) used for electricity generation, though the electric power sector gets only about 26% of its energy from coal. Electricity generation is an inefficient process as nearly two thirds (65%) of the energy supplied to the electric power sector is lost (23). There are many different models for managing the provision of electricity. The players include electricity producers, transmission and storage infrastructure, regulatory agencies, and consumers. In California, for example, an independent service operator (CAISO) coordinates the market for private electricity

generators, utilities that produce power and own transmission infrastructure, and consumers. CAISO does not own the generation or transmission assets but is a neutral entity that provides a transparent exchange for electricity supply and demand. CAISO does this by forecasting demand based on historical trends and allowing companies to buy and sell power before it is scheduled for delivery. Power is generated and distributed to utility substations and consumers through low-voltage lines based on the previous market run. CAISO also adjusts electricity flow based on supply and demand fluctuations in real time. In some instances, real-time increases in power demand, such as an increase in AC usage due to a heat wave, exceed projections and what the grid can provide, leading to "rolling blackouts" (34). A recent whitepaper from Southern California Edison, a major public utility servicing California, predicts major changes in electricity demand in California due to widespread adoption of electric vehicles. The paper projects a 60% increase in overall electricity demand and a 40% increase in peak load by 2045. Much of the increase will be more difficult to pinpoint geographically as electric vehicles are mobile and can be charged anywhere. Additionally, the increase in renewable power generators such as wind and solar will likely cause more variable power output as wind and solar capacity changes based on ambient conditions. Similarly, devices using inverters such as solar panels and batteries may also reduce the life span of electrical assets, potentially disrupting the continuous delivery of power (35).

Table 2.12 U.S. Energy Consumption by End Source, 2021

	Transportation	Industrial	Residential	Commercial	Electricity Power Sector
Petroleum	69%, 90%	25%, 34%	3%, 8%	2%, 10%	1%, 1%
Natural gas	3%, 4%	33%, 40%	15%, 42%	11%, 37%	37%, 32%
Renewable energy	12%, 5%	19%, 9%	7%, 7%	3%, 3%	59%, 19%
Coal		9%, 4%		<1%, <1%	90%, 26%
Nuclear	--	--	--	---	100%, 22%

Adapted from U.S. Energy Information Administration (EIA), "Monthly Energy Review, Table 1.3," April 2022, https://www.eia.gov/total-energy/data/monthly/.

Civilizations drink increasing amounts of energy as they develop technologically. The total U.S. electricity consumption in 2021 was nearly 4 trillion kilo-Watt hours (kWh), 13 times the usage in 1950, though the 2021 U.S. population was just over twice as large as 1950 (332 million versus 150 million). The U.S. demand is projected to increase significantly in the coming years with the transition to electric vehicles. Worldwide the biggest producers and consumers of energy are China and United States, with both countries leading in the production of both renewable and nonrenewable energy, as shown in Table 2.13. Though China (87.7 EJ or 24.4 trillion kWh in 2019, population 1.4 billion) is the biggest energy consumer in the world, the United States consumes far more energy on a per-capita basis (66.5 EJ, or 18.5 trillion kWh in 2019, population 330 million). The Chinese energy consumption was 1.3 times the United States', though the Chinese population was 4.3 times as large. The other top energy consumers are India, Russia, and Japan. Energy needs will increase most significantly in economically developing countries, especially those with population growth and urbanization in Asia and sub-Saharan Africa.

Table 2.13. Top Three Producers of Energy Sources, 2019

Energy Source	Top Three Producers
Crude oil	U.S. (17%, Russia (12,4%), Saudi Arabia (12.3%)
Natural gas	U.S. (23.6%), Russia (18%), Iran (5.9%)
Coal	China (49.7%), India (10%), Indonesia (7.4%)
Nuclear	U.S. (30.2%), France (14.3%), China (12.5%)
Hydroelectric	China (30.1%), Brazil (9.2%), Canada (8.8%)
Wind	China (28.4%), U.S. (20.9%), Germany (8.8%)
Solar	China (32.9%), U.S. (13.8%), Japan (10.1%)

International Energy Association (IEA), Key World Energy Statistics 2021, https://www.iea.org/reports/key-world-energy-statistics-2021/supply.

Energy-related activities are responsible for the bulk of GHG emissions worldwide. The respite offered by COVID-19 was short-lived. As the world economy rebounded in 2021, so did its appetite for energy. Global carbon dioxide emissions had fallen by 5.1% in 2020 but grown by 6% in 2021, mirroring the roughly 6% worldwide economic growth in 2021. Though coal-related emissions have gone down in the United States, they reached an all-time high in 2021, exceeding their 2014 peak by 200 MMT Eq of CO_2 (21). Electricity and heat production were the biggest drivers of CO_2 in 2021, accounting for nearly half (46%) of the GHG increase. China was the biggest source of electricity and heat-related GHGs, though China also pushed GHG emissions from its industrial sector below its 2019 levels (21). Unfortunately, coal-powered electricity plants were used to provide roughly half of the rising electricity demand for electricity in 2021. The carbon dioxide emissions from these plants exceeded their 2018 peak by 200 MMT CO_2 Eq. Though 2021 saw an increase in fossil fuel power electricity, renewable energy and nuclear energy combined to provide a larger share of the electricity produced worldwide than coal (21). In the United States, 81% of GHG emissions are due to energy-related activities, which include fossil fuel combustion and other activities related to the production, transmission, storage, and distribution of fossil fuels (5). Today, the United States accounts for about 14% of the 31.5 billion metric tons of CO_2 emitted due to fossil fuel consumption globally. Global energy consumption is projected to increase by 50% in the next 30 years if growth and consumption patterns continue unabated.

Food

Every living organism spends a considerable amount of time searching for food, or energy. This principle also dictated human existence until 10,000 to 12,000 years ago when the first human communities started cultivating wheat, barley, pulses, and other crops. The advent of agriculture transformed human society, allowing populations to grow and eventually paved the way for cities.[1] No longer bound by basic life-sustaining activities, many started to innovate in other areas. New fields and specialists emerged, further differentiating society. Hierarchies were established, as were new ways of living. Humans became less mobile, more specialized and interdependent, and technologies came to infiltrate all aspects of daily life. Art and culture flourished, and, over time, more egregious forms of oppression such as slavery were legally abolished. Technology made life easier and created new ways for humans to hurt and kill each other. Nowhere was this "double-edged" nature of technology more evident than in food production. The **Haber–Bosch process** for synthesizing ammonia (NH_3) from nitrogen and hydrogen illustrates how scientific progress and resulting technologies can be both savior and destroyer. Plants need nitrogen for a variety of life-sustaining functions and compounds, including photosynthesis, amino acids, proteins, and nucleic acids. Though nitrogen is abundant in the earth's atmosphere, it cannot be used by plants unless it is in a water-soluble form such as ammonia (NH^3). The Haber–Bosch process helps produce nearly 250 million tons of ammonia annually, which feeds roughly half of the world's population (12). The population growth during the 20th century, which added over 5 billion people to the planet, would not be possible without a process to extract nitrogen from the atmosphere. By some estimates, half of the cells in the human body contain nitrogen created from the Haber–Bosch process. The same process, however, also provides nitrates, a key component of explosives. During WWI, the Haber–Bosch process was used to manufacture explosives by the Germans as the naturally occurring deposits of nitrites in Chile were under Allied control. The process prolonged WWI, leading to thousands of deaths; the Nazis also used ammonia to neutralize hydrogen cyanide, which was used to murder millions of Jews in gas chambers.

The largest producers of ammonia today are China, Russia, and the United States. The energy used to power the Haber–Bosch process releases significant amounts of GHGs as almost all (98%) ammonia plants worldwide use natural gas or coal (11). In the United States, ammonia production accounts for 6.5% of natural gas consumed in the industrial sector; for every ton of ammonia produced, a U.S. plant releases on average 2.1 tons of carbon dioxide (11). In addition to climate change impacts, ammonia in agricultural water runoff can trigger **harmful algae blooms** (HAB) that kill marine life by depleting oxygen in waterways and blocking sunlight (13).

Food-related activities—production, processing, transportation, consumption, and waste—illustrate the multiple tensions between the different dimensions of sustainability. Humans require between 1,600 kCa–3,000 kCal daily depending on a number of factors, including age, weight, height, gender, hormone levels, and level of activity (9). Furthermore, the calories need to come from an appropriate distribution of **macronutrients** or carbohydrates, proteins, and fats. In addition to providing macronutrients, food must provide essential **micronutrients,** or vitamins and minerals, that cannot be made by the body as it can only produce vitamin D. Though human diets vary considerably across the planet, populations in different regions are increasingly reliant on processed and fast foods, especially in urban areas. Also, as people become wealthier, they tend to eat more **animal sources foods** (ASF), or meat. These changes in diet—more meat, processed and fast foods, larger portion sizes—during the last half century are implicated in heart disease, stroke, diabetes, and cancers, the leading causes of morbidity (sickness) and mortality (death) in the world today. Food and exercise choices, sometimes called lifestyle factors, contribute significantly to the world-

---

1. This is a simplified and teleological version of human history. In reality, the road to agriculture, hierarchy, and cities was neither linear nor smooth. Humans have experimented with many different ways of living, often choosing to revert to earlier forms of food procurement such as foraging despite having knowledge about agriculture. Similarly, cities were not always used for trade and commerce, and many societies did not develop strong hierarchical systems.

wide increase in **body mass index (BMI)**, a weight-to-height ratio Weights (Kg) / Height in Meters squared. It is estimated that today nearly 2 billion people are either overweight or obese (14).

Food production and consumption are subject to the same market forces that drive supply and demand for other consumer items. The Haber–Bosch process, which increased nitrogen supply for fertilizers, rapidly increased food production. The first **"green revolutions"** occurred between in the 1960s through 80s and used fertilizers, pesticides, new irrigation techniques, and new crop varieties with higher yields, less time to maturity, and greater disease resistance to substantially increase productivity per area (or hectare). The greatest impact was seen in Asian countries, where new varieties of wheat and then rice, such as IR8, were rapidly adopted. By adding fertilizer and pesticides to the **high-yield varieties (HYV)**, farmers were able to increase food supply by 12% to 13% between 1960 and 1990 (15). Food production has continued to increase during the past 2 decades: 9.4 billion tons of primary crops were produced in 2019, a 53% increase since 2000 (16). Sugar cane, maize, rice, and wheat together account for half of the primary crops produced throughout the planet (16). Meat production has also increased substantially during the last 2 decades: 337 million tons of meat were produced in 2019, a 44% increase since 2000, when 103 million tons were produced (16). The supply of milk (+52%), eggs (+63%) and fishery and aquaculture products (+41%) also grew during this period. The value of global food exports grew almost four-fold from $380 billion to $1.4 trillion in nominal terms (16). The increase in food supply has resulted in a 2,950 kCal per-capita per day **dietary energy supply** (DES) average. In Europe and North America, over 3,500 kCal per-capita per day are available. Similarly, protein availability is at an all-time high with a world average of 80 grams per capita per day. Given the high per-capita availability of calories and protein, it is no surprise that average dietary supply adequacy (12), which is a ratio of calories available to calories required, is at an all-time high (16). The impact on human health is evident. For example, the increase in food production has helped reduce stunting among children under 5 from 33% in 2000 to 22% in 2020. On the other end, the excess in calories has contributed to rising rates of overweight and obesity in almost all countries.

Though food production has increased steadily during the last two decades, progress has been impeded by a number of factors including poor governance, macroeconomic conditions, climate change, war, poor soil management, water depletion, cash crop production and, most recently, the economic impacts related to COVID-19. Those who are low income in developing countries, not surprisingly, were the most impacted. Between 80–110 million people were thrust back into extreme poverty in 2020, and 3.1 billion, 112 million more than in 2019, could not afford a healthy diet (60). The number of undernourished grew to between 700–820 million in 2021 (60). The largest proportion of the undernourished live in sub-Saharan Africa (20.2%), whereas the largest number reside in Asia (9.1%), specifically India. Worldwide, 8.6% were food insecure in Latin America and the Caribbean, 5.8 % in Oceania, and less the 3% in North America and Europe (60). In 2020, over 900 million people were estimated to experience severe food insecurity, and another 1.4 billion experienced moderate food insecurity. Almost a third of the world's population (30.4%) experienced either moderate or severe food insecurity in 2020; food insecurity remained relatively unchanged in 2021 (60).

Worldwide, obesity prevalence rose from 8.7% in 2000 to 13.1% in 2016. In the United States, 42% of adults were considered obese in 2020, an increase of 11% in the last 15–20 years (61). Obesity and overweight are defined with respect to BMI, calculated by dividing weight in kilograms by height in meters squared. A BMI of 18.5–24.9 is considered normal, 25–29.9 is overweight, and over 30 is clinically obese. BMIs in the overweight and obese ranges are correlated with adverse health conditions such as heart disease, diabetes, cancers, and premature mortality. Food, particularly processed foods, and physical activity play an important role in the onset of high BMI and related health issues (61). However, food and physical activity choices are conditioned by a number of biological, social, and political factors, many outside the control of individuals. The factors impacting food and exercise choices include the decreasing time available to prepare or cook food given work and other demands, increasing availability of processed and fast foods, greater market dependence for food, an increasing preference for animal source foods as incomes rise, reduced time and space for exercise, and auto-dependence. Highly processed foods often lead to overconsumption as they don't give the same satiety signals, or signals that the body is full, as whole, unprocessed foods; increasing portion sizes also encourages overconsumption (10).

The U.S. government enables food production in different ways, including prices supports for certain foods such as sugar and milk; production and market quotas; important restrictions, including tariffs; tax-related benefits; land and water related subsidies; and marketing and promotion programs (10). Lax regulations that allow food producers to use ambiguous marketing phrases such as "eat in moderation," "drink responsibly," "low in saturated fat," or "heart healthy" constitute de facto subsidies as companies are allowed to make claims to downplay health hazards or highlight health claims for which there is little empirical support.

Zoning in the United States also contributes to increasing market dependence for food and reduced mobility. Zoning separates land use, creating distances between residential areas and other land uses such as commercial, retail, and industrial. This separation increases auto-dependence and reduces active transportation such as walking or bicycling. Additionally, zoning discourages local food production by limiting the amount of food crops that can be grown in private residences or public lands. This limits the potential for local, barter-based food economies and compels people to rely on food markets, particularly large chain grocery stores. In many low SES areas, the only foods available are low-cost processed or fast foods, creating a cycle of dependence starting in early childhood that results in greater overweight and obesity rates in poor communities. This health inequality is compounded by the relatively lower access to quality preventive and primary care health care among the poor.

Waste

Human activity today generates a tremendous amount of waste. Like resource consumption, per-capita waste production tends to increase with rising income and urbanization. Higher income countries produce a lot more total and per-capita waste than lower income countries and account for one third of the world's total waste (48, 49). The worse offender is the United States, which is home to 4% of the world's population but produces 12% of the world's municipal solid waste (MSW). The United States also has the worst MSW recycling rates of any developed country at 35% (53); Germany, in contrast, recycles 68% of its MSW. Nonmunicipal sources of waste include agriculture, mining, and industrial processes. Though nonmunicipal waste often vastly exceeds the amount (weight) of MSW, much of it is handled in situ, *or where it is produced.* Waste disposal practices for non-MSW waste can result in environmental contamination. For example, agricultural waste, which includes pesticides, animal wastes, on-site medical wastes, and food-processing wastes, can lead to significant air, water, and soil pollution (54).

In 2020, an estimated 2.24 billion metric tons (about 5 trillion pounds) of waste was generated worldwide. This is equivalent to 1.7 pounds per capita (per person) daily. As shown in Table 2.14, about a third of the waste is dumped openly (48, 49). About 5% of total GHG emissions in 2016 were estimated to be related to open dumping and landfill disposal (50).

Table 2.14. Waste Disposal Worldwide

Waste Disposal	Percentage of Total Waste
Landfill Sanitary landfills w/ gas collection systems	37% 8%
Recycled or composted	19%
Incinerated	11%
Open dumping	33%

Adapted from What a Waste 2.0: A Global Snapshot of Solid Waste Management to 2050, by S. Kaza et al., 2018, https://openknowl-edge.worldbank.org/handle/10986/30317.

By midcentury (2050), we are expected to produce between 3.4 and 4 billion metric tons (8.6 trillion pounds), or about 1.1 kg or 2.3 pounds per capita daily. Much of this increase is going to come from the population and economic growth in developing countries, which are expected to double or triple their waste generation by 2050 (75). Like water management, waste management has serious implications for the environment and human health. In developing countries, urban slums are often located near trash dumps, and the urban poor sometimes supplement their incomes by scavenging items such as copper, aluminum, tin, glass, plastic, and paper from the dumps and reselling them (51). These activities increase health risks considerably, as trash contains hazardous materials and those living near the heaps are exposed to noxious vapors and gases, including methane. Children are particularly vulnerable as they are exposed to disease-causing pathogens, which also exacerbate stunting resulting from malnutrition.

As shown in Table 2.15, the United States and OECD-Europe countries already produce more than 2.3 pounds per capita of municipal solid daily. **MSW** includes waste from all sources (residential, industrial, commercial, etc.) collected by a municipality and disposed of at waste sites. Almost all countries in Africa produce less than 1 kg (2.2 lbs) per capita of MSW daily, as do China and India, the world's largest countries by population.

Table 2.15. Waste Generation (Total and per capita Daily) in the United States, China, Japan, and the EU

Country	Metric Tons (x1000) per Year	Population 2017	Pounds per Person per Year	Pounds per Person per Day
United States	243,724	329.8	1,625.8	4.5
China	215,209	1,410.0	335.8	0.9
Japan	42,894	126.7	744.8	2.0
OECD-Europe	283,182	512.0	1,216.8	3.3

Adapted from Organization for Economic Cooperation and Development, "Municipal Waste, Generation and Treatment: Municipal Waste Generated Per Capita," https://stats.oecd.org/Index.aspx?DataSetCode=MUNW.

Table 2.16 provides the breakdown by material of the 292 million tons (4.9 lbs per capita daily) of MSW pro-duced in the United States in 2018. Over 64 million tons of food was wasted in 2018. This represents about **1 pound per capita of food wasted** daily, of which half is usable (200 lbs per person per year). Of the total MSW produced, 24% (69 million tons) was recycled. The largest category of recycled materials was paper and paperboard. Plastics recycling rates were an abysmal 4.5% (52). Plastics can make their way to marine life through gyres that circulate water in the oceans. Plastics cause harm by releasing toxic chemicals, causing entanglement and ingestion and dis-tributing non-native and harmful organisms (55). The plastic concentrations in some subtropical gyres are so high (hundreds of kg per Km^2 or > 1 million pieces per Km^2) that they are colloquially referred to as a "garbage patch" (56). Humans also ingest microplastics daily, the long term impact of which is likely quite harmful but still poorly understood. It's possible that microplastics in humans may have an effect similar to asbestos or chemical toxicity (57, 58).

Table 2.16. MSW and Recycling Rate in the United States (2018) by Material

Material	% of Total MSW (292 Million Tons)	% of 69 Million Tons Recy-cled
Paper and paperboard	23%	67%
Food	22%	
Yard trimmings	12%	
Plastics	12%	4.5%
Glass and metals	13%	17%
Wood	6%	4.5%
Other (rubber, textiles, inorganic, misc.)	12%	7%

Adapted from EPA, "National Overview: Facts and Figures on Materials, Wastes and Recycling," https://www.epa.gov/facts-and-fig-ures-about-materials-waste-and-recycling/national-overview-facts-and-figures-materials.

Waste management is the least cost effective at the end stage when waste is deposited into landfills when any residual economic value is lost. As in public health, prevention is the ideal strategy; however, this would reduce the number of consumable goods, impacting the economy. After prevention, recycling and recovery allow materials to be repurposed or reused. These strategies are often expensive and are not implemented until regulation compels changes in the marketplace. Some countries in the EU and Japan have made significant strides in diverting waste from land-fills to recovery and recycling in the last 2 decades through legislation, regulation, and market-based incentives (49). Developing countries also need to invest heavily in waste management to divert waste from open dumping to regu-lated facilities. However, waste management is expensive, as development and operational costs are high. In higher income countries, **integrated waste management**, which includes collection, transport, treatment, and disposal, is $100 or more per ton; lower income countries spend about $35 per ton (75).

Conclusion

The preceding discussion illustrates the various ways human civilization impacts the environment. The road to economic growth often exacts a health cost that is carried disproportionately by marginalized populations such as the urban poor. Economic growth has lifted many out of poverty in the last few decades; however, its fruits have not been shared equally. Growth also demands much from the environment as resource consumption, waste, and GHGs alter the earth's ecologies in irreversible ways, causing significant and lasting harm to both humans and nonhuman organisms. Some, tragically, are pushed to extinction and will never inhabit the earth again.

In response, the public sector has enacted legislation and regulations. In the United States, important environmental regulations and legislation include the creation of the Environmental Protection Agency (EPA), various pollution-related acts (e.g., Clean Air and Water Acts) and pollution control programs (e.g., NPDES). Other countries have done the same and have created agencies for environmental oversight and enacted regulations that govern everything from recycling rates to GHG emission. Many have reversed their environmental footprint in certain areas, and others, like the United States, continue to lag in areas such as waste management. Governments have also provided incentives such as tax credits for electric vehicles to force a pro-environmental market reaction. Some companies such as Tesla have created automobiles that are desired by consumers; legacy auto manufacturers have finally joined the fray, recognizing shifting consumer demand, at least partly due to higher fuel prices, and in response to incentives and regulation. In construction, LEED and other related standards have been adopted to increase the sustainability of the built environment and reduce its environmental footprint. Other parts of the private sector, recognizing the "cultural moment," have stepped up, at least on paper, and rebranded themselves as a "green business." Some have explicitly incorporated sustainability into their corporate DNA through some type of ESG (environmental and social governance) framework (others are compelled by large investors such as Blackrock or a regulatory entity). Other legislation incentivizes GHG reduction by creating carbon credit markets that allows private companies to trade carbon credits. International environmental agreements, the most significant of which was the 2015 Paris Accords, have been signed by countries to curb waste, limit resource consumption, and address climate change. The Paris Accords were hailed by some as an important step toward limiting the increase in global mean temperatures to below 2 degrees C (3.6 degrees F), whereas others criticized them for lacking "teeth" (enforceability). Some issues are proving to be more intractable: water scarcity, which is exacerbated by climate change, is likely going to worsen as water use increases with economic growth, though the full cost of water is difficult to pass onto consumers as it is regarded as a basic human right. In richer nations, even the most conscientious persons—those who conserve water diligently at home—will use a lot of water indirectly because of the food they eat or clothes they wear.[2] Much progress has been made, but much more remains to be done. In many ways the problems are inherent in the global-capitalist mode of production, and all efforts to address the various environmental issues are piecemeal and bandages

---

2. Fast Fashion has increased the per capita consumption of clothes throughout the planet. The environmental toll with respect to resource consumption (e.g. cotton, water, energy) and waste is considerable. However, on the flip side, industries related to fast fashion employ millions, though many work in sweat shop like conditions.

at best. The tensions between economic well-being and the environment are hard to reconcile, particularly in developing nations that also want a piece of the "good life" and simply don't have the resources and capacity to take the steps demanded of them by richer nations that have polluted without shame or consequence for decades prior. The richer nations are often unwilling to pay for their past sins. Time will tell if the world is truly able to come together to address the multiple threats to the collective well-being of all inhabitants, human and otherwise, of this planet.

Chapter 3

Urbanization and Globalization, a Historical Perspective

Key Terms

Trade: The exchange of manufactured goods, commodities, cultural items (art, pottery), and services for money. Trade can also involve barter, which is the exchange of these aforementioned items for other items without using money. Trade has been an integral part of the world economy since the first human civilizations. Trade can help increase the material standard of living in a region by increasing the amount and quality of goods and services available in the region. Trade can also help increase cultural and social exchange between regions, increasing understanding and cooperation.

Colonization: The military, political, economic, and social control of one region by another. Colonization, in common usage, refers to the conquest and control of regions in Africa, the Americas, and Asia by European states. An impact of colonization was **mercantilism**, an economic system in which the colonizers take resources, usually by force or coercion, and transform those resources into finished goods that are exported back to the colonies and other parts of the world. This process enriches the colonizers as they don't have to pay the true cost of those resources.

Integration: This is an important feature of modern globalization. It refers to the depth of linkages and interdependencies between regions throughout the world. The term typically focuses on economic integration or the trade networks between regions and capital integration, or the flow of money through public and private investments between regions. Integration can also refer to social, cultural, and population (i.e., emigration/immigration) exchange between regions.

Other key terms will be in **bold** and defined in the text.

Review Questions

- How is globalization conceptualized? What is glocalization?
- What is the Enlightenment, and what are its core tenets?
- What is modernity?

- What is dependency theory?
- What were the characteristics of the first urban areas, including trade networks?
- How did the Roman Empire enable trade and globalization?
- How did the Mongols enable trade and globalization?
- How did Europe gain economic power? What is mercantilism?
- What is industrialization, and why was it disruptive?
- What role did technology play in globalization in the 19th century onward?
- What are the factors that help explain the ascendancy of the United States?

Introduction

In the preceding chapter, globalization was described primarily as an economic system. However, there are other ways to think about globalization. If globalization is more than an economic phenomenon, what are some of its other central, or primary, features, and what are its secondary characteristics or the effects of its primary characteristics? Additionally, how did these features come to be, or what are their historical antecedents? Is there a **teleology** to civilization, an evolution that improves civilization toward some sort of ideal state? Conversely, what are the problems with looking at societal change in terms of a teleology that starts with hunter-gatherer societies and culminates in modern, democratic nation-states connected by global capitalism? There are some clear historical markers that presaged the type of globalization we see today; however, the historical record also contains a remarkable number of examples of societies that did not simply progress in a stepwise fashion from foraging to industrialization. Cities were built as sacred sites for noneconomic purposes, and many did not have a ruling class (e.g., Jomon in Japan, Teotihuacan in Mesoamerica); trade occurred for noneconomic goods (e.g., the heirloom arm-shells and necklaces traded by men in the Massim Islands to spread their name); a strong ruling class or political hierarchy did not always accompany specialization and a high degree of urbanization (e.g., Indus Valley); some cultures had leaders only during war or leaders with very specific purpose and limited coercive power; property rights were consciously rejected; and farming was not embraced despite an intimate understanding of how to farm.

Though today urbanization, specialization, globalization and democracy (at least in name) seem predestined given that almost all societies have adopted them in some measure, these preceding examples show that human societies have experimented with a bewildering number of **ontologies**, or ways of being. It is our capacity to imagine and organize in many different ways that give us our most powerful weapon against the dangers of our current civilizational path. The first step in changing is believing that change is indeed possible. We must first understand modernity and its many children, which include positivism, democracy, and global capitalism, as simply one of many possible paths. Modernity offers a specific **epistemology**, or way of understanding the world, and legitimizes certain ontologies, namely those that it casts as "rational." The abundance of material wealth created by industrial global capitalism and the nexus between democracy and improvements in human rights have led many, including policymakers, to argue that all societies should adopt these models. However, there are other ways of knowing the world and different ways of being that don't fit neatly within the fold of modernity. If we imagine that change is possible and desirable, we can begin to experiment anew. This does not have to lead to a complete rejection of modernity. However, we have to be willing to discard the things that are destructive. To do this, we must first identify what those things are. This is difficult as it requires examining the things we take as given and, potentially, choosing alternatives that make life less convenient but are better for us and the world. Michel Foucault, a philosopher, argued that modernity's various children (businesses, government, technical experts) assume the cloak of reason, impartiality, expertise, and compassion but, in reality, subjugate humans to various systems of power. Paradoxically, today, humans are more free on paper and able to access a plethora of goods and services, but few can survive outside the market and the other systems of power that exert control almost every domain of life (e.g., formalized systems of law; police power of the state [i.e., regulations]; specialists in every domain, from dating to pet care; large corporations that enable the good life). In the following sections, we look at the conventional explanations about how modern society is and came to be. This narrative is not wholly incorrect, but it elides important exceptions and leaves little room for nuance. It presents itself as the only answer. In the last chapter, we return to the idea of imagining alternatives and what you, as an individual, can do.

Urbanization, specialization, and globalization are intimately linked. Urbanization was first possible because of surplus food production. Specialization arose in urban environments, improving the quantity and quality of goods produced. Local trade eventually expanded into regional trade. Trade fostered economic, cultural, and social exchange, making regions more interdependent though nowhere near to the extent to which they are today. This sometimes led to cooperation across a number of domains, prosperity, and peace. The first city-states in Mesopotamia and Indus Valley, though separated by 2,500 km, engaged in trade over land and sea. In addition to exchanging goods, artifacts, and commodities, they influenced each other culturally. This pattern would be repeated throughout history and continue today with exchange occurring over real (physical) and virtual (internet) networks.

Specialization and urbanization played a critical role in the industrial revolution, which improved production processes, substantially improving product quantity and quality. New types of jobs were created, and incomes increased. However, many jobs become obsolete, and resource extraction and environmental degradation, including climate change, increased substantially. As societies became richer, the spoils (income and wealth) were not shared equally, with those at the top taking an increasingly larger share of the newly generated income and wealth. These issues notwithstanding, the collapse of the USSR (Soviet Union) in 1991 was supposed to herald a new age for humanity, one characterized by global capitalism, democracy, increasing prosperity, and human rights. However, the last 30 years have seen a global recession (2007–2008), the rise of nationalism throughout the planet, a pandemic, and, most recently, the invasion of Ukraine by Russia and a major military action by Israel against Hamas in response to a brutal incursion by the group into Israel on Oct 7th that has the potential to engulf both regional and other actors like the US. All of these events underscore the fragility of the global world order and the hubris of those who believed globalization to be, more or less, a completed project.

This chapter explores the many different frameworks used to understand globalization, including world-systems theory. The chapter also examines specialization, trade, urbanization, and integration historically to better understand globalization today and its precursors and possible trajectories. By looking at the history of trade, exchange, and technological development, we can better understand our modern global system and where it may be headed. Our story begins with the first urban civilizations in Mesopotamia and Indus Valley and goes through the ancient world, including the Roman Empire, the Mongol Empire, and the silk routes, the largest trade network of the ancient world. The chapter then examines the beginnings of the modern world, starting with the discovery of the Americas and colonization by the European powers. The chapter examines the impact of the three waves of industrialization (mechanical, electrical, and computerization) and their impacts and discusses the implications of the fourth industrial age being forged today. We end our journey by discussing the impact of the two World Wars on globalization and the rise of the United States.

The Many Definitions of Globalization

Globalization has become a catchall phrase that encompasses many different ideas. Some use the word to indicate the spread of Western, specifically American, values, such as capitalism, individualism, and democracy. In this sense globalization is synonymous with **modernity** and represents the triumph **of enlightenment** values. Others see globalization as a technologically enabled force that **homogenizes** the world through the exchange of goods (i.e., trade), ideas (i.e., media and internet), and people (i.e., immigration and travel). In this view, people increasingly consume the same products, eat the same foods (often fast foods), indulge in the same types of entertainment, albeit in different languages, and believe in the same things; consumption plays an extremely important role in this perspective. Many see globalization in mostly economic terms, specifically the spread of market-based economies, **deregulation, and privatization**. Others argue that globalization is a political phenomenon. They see globalization in terms of the world order that emerged after the end of the Cold War (1). Some argue that "real" globalization is a type of **integration** that can only occur in the modern era through technology, specifically the internet and transoceanic transportation, which allows the various parts of a product's **supply chain** (i.e., where the different parts of a product are made) and **capital**, or financing for the product, to be located in different regions. This also allows for the flourishing of transnational or multinational corporations, with production spread across the world.

Glocalization is a related concept that attempts to capture the tension between the old order, in which companies and countries were inextricably linked, and contemporary transnationalism in which companies have little or no patriotic inclinations. **Glocalized** companies will locate production abroad to curry local favor and avoid punitive actions but still maintain corporate control and the bulk of their R&D facilities in their country of origin (1). The more optimistic integrationists see globalization as an ineluctable or unstoppable, postindustrial force that results in a single world market in which buyers and sellers for all types of goods and services are free to move around without national or other constraints. In this view there is a fundamental reorientation of the world economy away from nation-states and the old industrial order toward a more integrated, interdependent, and free world.

The fervor surrounding cryptocurrencies as an alternative to traditional sovereign-backed currencies, particularly the U.S. dollar, can be seen as one expression of this perspective. There are other less sanguine, or less optimistic, views. These cast globalization primarily in terms of its **externalities**, or ill effects, specifically the increasing market dependency of the masses, growth in consumer culture, wage and wealth inequality, and the growth of stateless corporations that perpetuate and profit from the preceding trends. These globalization naysayers also highlight the environmental and climate impacts of increasing resource extraction and waste and, in a nod to Durkheim and Marx, the growing alienation of humans from themselves and others. The alienation is the result of market dependency for all goods and services, virtual work and entertainment, and entrenchment of individual ethos or hyperprivacy, all of which reduce **social capital**, the informal ties that serve as the glue of communities. Others emphasize the relationship between globalization and disease and conflict; the COVID-19 pandemic is in many ways a story of globalization: The rapidity of its spread, the development of vaccines, and resistance to public health measures were all enabled by the forces of globalization. The Russian incursion into Ukraine and its fallout have significant globalization dimensions: An interconnected world means democracies will sometimes rely on authoritarian regimes such as Russia, which is a large energy (natural gas, oil, and coal) and agricultural exporter. Despite sanctions on its oil by major Western powers, Russia has still managed to rake in billions in oil revenue (34). Ukraine, like Russia, is an exporter of agricultural goods such as seed oil (e.g., sunflower oil), wheat, corn, and fertilizer. A protracted war is expected to reduce economic growth among their largest trading partners, many of which are in Asia and Africa (35).

Other critiques come from a world systems perspective and **dependency theory**, which examines core and periphery relations such as those between industrialized nations and developing countries. A world systems perspective argues that "surplus value" is expropriated from developing nations by developed nations in the global-capitalist order. World systems ideas, however, decenter the nation-state by emphasizing the importance of transnational forces such as the capitalist class or large corporations that shape the development trajectories of regions, including countries. In this view, countries themselves are not independent agents but subject to forces that shape relationships between countries and within countries. Additionally, the perspective calls into question that idea that there is only one path to development (9, 11)

Mesopotamia and Indus Valley

The first civilizations emerged in multiple locations throughout the world. These areas, sometimes called the "cradles of human civilization," shared several common features, including urban development, government, social hierarchy, agricultural surplus, craft specialization, symbolic communication (language, art), and trade. The earliest of these civilizations were located in regions that are largely coterminous with the countries of Iraq, Pakistan, Egypt, and China.

Humanity's first major cities arose between the Tigris and Euphrates Rivers in an area that is also known as Mesopotamia, or the land between rivers. This region was the birthplace of many of the appurtenances we associate with human civilization. For example, the wheel and writing were likely first developed here. Similarly, one of the first instances of codified law, the Laws of Hammurabi, can be traced to this area. Starting with the Sumerians in the 6th millennium BC, many different groups, including the Akkadians, Hittites, Babylonians, and Assyrians, would carve out their legacies by building cities, creating distinct cultures, and engaging in both warfare and trade.

The domestication of crops, large-scale agriculture, and better water management, including irrigation, allowed for specialization in food production and activities unrelated to food production. Though food production was still the biggest economic activity in early Mesopotamia, there was a seeping specialization, even among farmers and those engaged in animal husbandry as some exclusively grew grains, others herded, some hunted, and others became fishermen. Improvements in agricultural yields due to specialization and innovation enabled food surpluses, which freed many from the tasks associated with food and water procurement, allowing them to focus their energy on other endeavors. This specialization in food production was an important precursor to craft specialization and the production of textiles, pottery, jewelry, woolen goods, and other items for trade between the cities in Mesopotamia, as well as trade between Mesopotamia and other civilizations, such as those in the Indus Valley or Egypt. The domestication of crops thus allowed for the emergence and flourishing of trade. Additionally, specialization wasn't limited to those engaged in some type of craftsmanship. Many also focused on other areas, such as writing, law, medicine, and religion, necessary for the functioning of large, complex societies.

Specialization increased the speed with which products were produced and their quality. Newer technologies and innovations allowed for even more improvement. Trade flourished within Mesopotamia between 3000 and 1500 BCE. Trade networks became more complex and longer, and regional specialization increased interdependency between city-states (2, 3). Technological growth accompanied growth in trade, as did improvements in transportation. These factors likely exerted a reciprocal impact on each other, creating a positive feedback loop. Though the exact factors leading to an increase in trade are not fully understood, it is likely that growth and specialization in food production and merchant institutions played an important role in fostering trade (4). Specialization in food production increases the volume of food produced with the surplus potentially available for trade. This allows spatially proximate regions or regions in a trade network to share risks related to food as these regions could rely on each other for food in times of distress. This strategy also allows for further growth and innovation in areas unrelated to food production as resources were directed to other nonfood-related areas.

The Indus Valley city-states, another major early civilization located in modern-day Pakistan and Northwest India, were different than their Mesopotamian counterparts in some important respects. The cities in Indus Valley occupied a much greater geographical area, and their population was considerably larger (8). Urban design characteristics were relatively more unform across Indus cities as well. For example, brick sizes were standardized, as were street widths, and there appeared to be a rudimentary form of zoning with different areas of the cities used for distinct purposes. Indus cities also had advanced water management systems that included gravity drains and sewage. Many cities had granaries, large baths, citadels, and dockyards. However, there were no large monumental structures, such as the Mesopotamian ziggurat. Similarly, there is little evidence of a strong military culture and related technologies. The social and political organization of the Indus city-states is still far from clear as the Indus language is yet undeciphered. These features have led some scholars to speculate that Indus Valley did not have a strong social and political hierarchy like their Mesopotamian neighbors, though this is far from certain.

A considerably body of research has established the existence of a trade network between the Indus and Mesopotamian cities (5, 6). Scholars have found both objects of Indus origin in Mesopotamian sites as well as genetic evidence linking the two areas (6, 7). A larger number of Indus artifacts, such as beads, seals, weights, figurines, and pottery, have been found in Mesopotamian sites, and Mesopotamian texts refer to Indus traders as those from Meluhha (6). Mesopotamian writings suggest that woolen items, gold, and other perishable goods were traded in exchange for Indus finished goods. The distance between the two areas was considerable. Merchants had to travel 2,500 km through land or sea routes to ply their wares. This distance required the creation of trade networks and trading colonies along the trading routes.

As trade flourished, so did the wealth of the urban areas involved in trade. The growth of cities also increased conflict, which was the impetus for a refinement of the tools (i.e., weapons) and roles (i.e., military hierarchy and organization) necessary for engaging in more sophisticated types of warfare. Specialization thus greatly improved the capacity of city-states to wage war by supplying them with better weapons and people trained specifically to function as warriors.

This development could be considered a nascent or proto- military-industrial complex. Unlike today, the first city-states did not have standing armies comprised of those specializing in war. In some cities on a war footing, all free male residents were expected to fight and to bring their own arms to battle. Over time, however, a professional military became *de rigueur* for those ruling the city-states, some of which became part of a larger federation, kingdom, or empire. The inclusion of the conquered into the forces of the conqueror also became standard practice. The *casus belli* were usually not much different than wars of recent memory: Annexation of additional land, resources, retaliation, and differences in ideology were all used as justifications.

As cities grew, so did the toll exacted upon the natural environment through agricultural intensification, resource extraction, and waste production. The earliest urban centers did not produce anywhere near the per-capita waste as communities today. However, their impact on the environment was undeniable, certainly when compared to their nonurban ancestors. Southern Mesopotamia, home to the oldest urban centers, had more rain fall during the spring, though their crops required the most water during the fall. To ensure a steady water supply during the fall, water storage and irrigation became paramount. Though water storage was instrumental in securing a steady food supply, over time this increased the salinity of the soil. Waterlogging caused by low soil permeability exacerbated by the silt deposited by the rivers increased the amount of water in the soil (i.e., the water table), which left greater amounts of salt as it evaporated in the summer months. The increasing soil salinity in Sumerian cities due to agricultural intensification was thus a major reason for their decline and the shift in power northward to the Babylonians. Similarly, the inability to address the impacts of naturally occurring climate change, which caused a reduction in rainfall, has been implicated in the decline of Indus Valley (Harrapan). The relationship between urbanization and environmental degradation is a recurring theme in human civilization. Today the implications of this relationship are more consequential than ever given the size of the human population, resources needed to maintain a modern standard of living, and waste generation.

The Roman Empire

The history of classical Rome is well documented academically and explored in the popular imagination through print and audio-visual media, though these have often been fictionalized accounts developed primarily for entertainment. Historians often differentiate between the Roman Republic and the empire. Rome was unique in that democratic government preceded authoritarian rule. The republic, which formed in the 6th century BCE and was governed by an elected senate, lasted 500 years. Its dissolution occurred at the hands of Julius Caesar, arguably the most famous Roman emperor, who ascended to power using the clout and wealth generated by his many successful miliary campaigns, starting with his time as the governor of Gaul, or modern-day France. Caesar was assassinated, and his nephew Gaius Octavius ascended to the throne. Octavius was known as Augustus and ushered in a period known as the Pax Romana or Roman Peace (though it was also marked by conflict but to a lesser degree than the strife characteristic of earlier periods). For the next 200 years, Rome was ruled by the "Five Good Emperors," the last of which, Marcus Aurelius, is often known as the philosopher-king for his adherence to Stoic philosophy. The western empire crumbled in 476 AD as Rome itself was sacked by Odoacer, a Germanic chieftain. The Eastern Roman Empire remained under Zeno, and Odoacer shrewdly made himself a **satrap**, or lesser ruler, under Zeno, though in reality, he was the de facto ruler of the Western Roman Empire, which would later become the papal states, and the eastern empire would continue for nearly another millennia until it was overthrown by the Ottomans in 1453.

The legacy of the Roman Empire cannot be overstated. The Roman contribution in terms of technology, ideas, and culture to both Western civilization and the world was significant. A quick internet search of "Roman Empire contributions" yields a diverse set of results that include the Julian calendar; architecture, planning, and engineering advances such as the arch, aqueducts, concrete and the dome; a representative model of government; medical innovations; artistic traditions; and a Latin-based lingua franca that was the precursor to many contemporary European languages. Arguably, the greatest Roman contribution was facilitating the spread of Christianity. Though initially a religion practiced by no more than a small proportion (3% to 5%) of the Roman population, Christianity became the official religion of the empire under Theodosius in 380 AD. Earlier, Constantine had declared, through the Edict of Milan, the legality of Christianity and other religions practiced in the empire. Constantine also convened the Council of Nicaea in 325 CE to help establish the formal theological basis for Christianity.

The extent to which the Roman world was globalized is still a hotly contested topic among scholars, with some arguing that modern scholars diminish the globalized nature of the Roman era and others making the case for a less Roman-centric view of globalization during antiquity given the presence of trade networks along the Mediterranean and beyond before the republic and Pax Romana. The arguments are, to some extent, academic as none denies the influence and impact of the empire in bridging far-flung areas and enabling the exchange of goods, ideas, and people.

These impacts notwithstanding, the extent to which the Roman world constituted a truly globalized world system will likely continue to provide fertile ground for scholarly debate in the years to come.

Mongols and the Silk Route

As we saw in the preceding section, the Roman empire played a key role in creating the conditions for the economic and cultural integration of regions across a vast area that included large swaths of Europe, portions of North Africa, and even some parts of the Middle East. The Romans were explicit in their desire to create a world based on their ideals. Empires and their armies have historically provided the stability necessary for smoothly functioning trade networks, which help spread not only goods and services, but also facilitate the exchange of culture, worldviews, ideas, and diseases. The Romans controlled many of the port cities along the Persian Gulf, the Red Sea, and Mediterranean, allowing for a smooth flow of goods to and from the east. These trading networks supplied the Romans with silk from China and spices from India. Whereas the Chinese and Persians were sometimes hostile to traders, the Mongols, like the Romans, realized their importance and took steps to maintain the continuity of trade through regions under their control. The silk routes that connected China and Europe reached their height under the Mongol Empire, which not only provided protection for traders, but also took steps to ensure the smooth functioning of trade capital markets by making loans. In some cases, the Mongols destroyed smaller cities along alternate routes to streamline trade routes. It is hard to understate the impact of trade networks in shaping the future of Europe as it emerged from Medieval times (i.e., the Dark Ages). The Renaissance was a literal rebirth for Europe. At their vanguard were the Italian city-states, including Venice, which became prosperous through its trade with the east through the Mongol-controlled silk routes. Marco Polo and his family similarly benefitted from their partnership with the Mongols.

In terms of contiguous land, or land that is adjacent to another tract of land, there has never been an equal to the Mongol Empire. The Mongols were nomadic and depended on pastoral lands and water for their **ungulates** (e.g., goats, cows, sheep, horses). This lifestyle was inherently precarious and required frequent movement. In 1206, a young Mongol warrior names Chinggis (Genghis) Khan unified the Mongols and gave birth to the Mongol Empire. Recent research postulates that a milder climate and greater moisture may have made it easier to sustain horses and camps as grass and vegetation were more abundant during the unification campaign. The Mongols mastery over riding and archery allowed them to stage frequent raids and destroy opposing towns and armies. Though they were fearsome in battle, they proved even more able as administrators of a large land empire. Some 70 years after the Great Khan unified his people, there began a protracted period of peace and stability, known as the *Pax Mongolica*, that allowed trade to thrive along the silk routes.

The bidirectional trade along the silk routes was instrumental in the exchange of goods, culture, and knowledge during Mongol times. Traders and others using the silk routes greatly expanded European knowledge of civilizations along the route and vice versa. The age of exploration or discovery, which saw European seafaring civilizations spread out across the world, was in many ways the beneficiary of the knowledge gained through the silk routes. Though routes to the east had existed prior to *Pax Mongolica*, it was rare for a single traveler to make the entire east-to-west journey. Knowledge before the Mongol period was often fragmented, incomplete, and, in some cases, wholly incorrect. The Greek, for example, until the 6th century of the common era (CE), believed that silk came from trees. For much of its existence, trade along the silk routes was mostly fragmented as it had to pass through various subnetworks, intermediaries, and spheres of influence. In some cases, the intermediaries, such as the Parthians in Persia, discouraged direct connections between Europe and China to buttress their power. It's incorrect to speak of a single silk route or a cohesive network with established routes and rules. Rather, depending on the era, there were many different land and sea routes connecting Europe and China.

Many goods were exchanged along the silk route, but no less important was the cultural diffusion that occurred because of trade. Some of the effects were direct, such as use of Persian or Arab coins, whereas others occurred over time and were indirect (14). For example, Arab weight systems for precious metals were adapted by northern Europeans gradually. Similarly, Scandinavian cultural artifacts, weaponry, and clothing were influenced by the central Asian nomadic, steppe culture (14, 15). Even Buddhist and Indian iconography found its way into northern European artifacts through the silk routes (14,16). For roughly 1,600 years, from 200 BCE to 1400 CE, the silk routes provided a series of thoroughfares between and through major civilizations of the world who traded goods, ideas, and people with each other. One of their legacies, explored in the next section, was the European age of exploration or discovery, which opened up new trade routes, led to the discovery of the Americas, and laid the foundation for European colonialism and resurgence.

Age of Discovery and Mercantilism

European seaward expansion in the 15th century, the start of the age of discovery, that led to the colonization of the Americas and circumnavigation of Africa was not a forgone conclusion. Though Europe had a maritime culture, it was not the dominant seafaring civilization prior to the age of discovery. Medieval Christian Europe was largely relegated to the periphery after the fall of the Western Roman Empire. Scientific, technological, and cultural progress was stifled and, in many instances, thwarted by the Roman Catholic Church. Europe was not renowned for its arts, culture, technology, goods, or learning. The important seafaring cultures were mostly Eurasian and included China in the Pacific Ocean, India in the Indian Ocean, and parts of Southeast Asia, and the Islamic cultures in the Middle East and North Africa. The Chinese flotillas tended to be massive. In 1405, at roughly the same time Europeans were exploring the coast of West Africa, a Ming Dynasty's expedition along African's eastern coast consisted of 62 ships and nearly 28,000 men (17). This was a scale several times larger than what the European powers could assemble. Similarly, in the Americas, which were yet undiscovered, there were many cultures, some controlling tens of thousands of square miles of territory. As roughly the same time as the age of discovery, the Aztecs dominated Mesoamerica in the area around Lake Texcoco (modern-day Mexico City), and the Incan Empire, centered in modern-day Peru, dominated the western part of South America. Each reached its end at the hands of European colonizers who were aided in no small part by the new diseases they brought to the Americas.

In Europe, trade was monopolized by the Italian city-states, Genoa and Venice in particular. These cities became quite wealthy and used that wealth to support exploration as well as scholars, artists, sculptors, and the like; the Medici, a powerful family based in Florence, even funded a papal campaign and became the bankers of the Papacy. The Italian support for learning and arts, along with a rediscovery of classical knowledge through a study of stored classical manuscripts, gave birth to what we know as the Renaissance, which literally means "rebirth." The technical prowess of the Roman Empire had been lost for generations and would not be reclaimed until the Renaissance.

The Middle East, in contrast, had blossomed intellectually, culturally, and politically under Caliphates and Persian rulers. The Crusades had done little to weaken their ascendancy. It was not until the 15th century that conditions were ripe for a European rebirth. Though much was known about North Africa, the information on the rest of the continent was scant. Since the Italians controlled Mediterranean trade, there was a great desire on the part of the other European states to find alternative trade routes to the Middle East, South and Southeast Asia, and China. The collapse of the Mongol Empire made traveling by land more treacherous. Additionally, the depopulation of Europe due to the bubonic plague (or black death) in the 14th century and the rise of the Ottoman Empire, which deposed the Eastern Roman, or Byzantine, Empire, provided further fuel for exploration of alternative sea routes. The conditions were ripe in the early 15th century as political will, funding, and technological advances, specifically **the caravel**, a small ship that could sail windward and deeper into the ocean, allowed the Portuguese to explore the southern parts of Africa and the Atlantic Ocean with the aim of developing alternative routes to the Middle East and East Indies (India) to grab a portion of the lucrative spice trade. Christopher Columbus, who had first pitched his idea to travel westward on the Atlantic to the Portuguese, set sail in 1492 with the aim of reaching the Indies. Instead, Columbus reached the Americas and founded the first European colony in Haiti. Columbus forcefully took some natives back with him, of which only a fraction survived. However, they provided sufficient proof of the exotic potential of the new world.

The Chinese unification was both a blessing and not. A centralized state made it easier to direct resources more efficiently, but it also stifled competition. European disunity, in contrast, provided a fertile ground for competition. As the various European states vied with each other, they developed new tools, techniques, and technologies for exploration and then colonization (17). Many other cultures possessed similar attributes, but the Europeans standardized and institutionalized many of their advantages. The development of the printing press by Johannes Gutenberg in 1439 was also, undoubtedly, instrumental in the diffusion of knowledge throughout the continent. The discovery of the Americas by Columbus in 1492 gave the Europeans an alternate route to riches, as did the circumnavigation of the Cape of Good Hope by Vasco De Gama in 1497, which gave the Portuguese a new route around Africa to India.

By the 16th century, the European powers only had two real avenues for increasing their economic output. They could expand trade with Asia or exploit the new world. Asia was the safer bet; however, there was a greater upside to the Americas (18). They pursued both strategies simultaneously, with the Portuguese initially focusing on Asia and Spain, the new world. The Dutch supplanted the Spanish and Portuguese through the activities of the Dutch East India Company. However, by the end of the 16th century, a new player had emerged. The British soon came to dominate world trade through their superior navy. They became the biggest masters of the new world as well as the old through their colonization of India. Other European states also carved up Africa, Asia, and Southeast Asia. Colonization gave rise to **mercantilism**, the belief that the world economy was a zero-sum game that required accumulation of commodities, including slaves and bullion as well as limits on imports and an increase in exports. Some colonies such as India became **colonies of extraction**, in that European states expropriated or took valuable commodities (e.g., wool, cotton, and spices from India), and others became **colonies of settlement**, in that immigrants from Europe built new communities and, eventually, nations (19). The colonies of extraction were often more developed and had higher population densities than the colonies of settlement. However, over the long run the colonies of settlement were more inclusive, with greater property rights and incentives for entrepreneurship, largely because they were comprised of European settlers (19). Unsurprisingly the development trajectories of the two types of colonies were quite different, with colonies of extraction lagging far behind those of settlement.

These practices continued until the industrial revolution, which radically altered the productive capacities of economies and upended the older perspective in which the world economy was a **"zero sum game,"** a system in which a win required another to lose, or, alternatively, a world economy with finite riches: An increase in one state's share decreased another's.

The Industrial Revolutions

The industrial revolution first occurred in England. Many reasons are proffered as to why: the growth of finance, urbanization, and social and cultural changes; improvements in transportation infrastructure such as roads, canals, and later rail lines; the discovery of important natural resources; and population growth. There is also some debate around how "industrial revolution" is **operationalized** (i.e., what is meant by the phrase "industrial revolution"). These issues notwithstanding, the historical data show that England underwent a dramatic transformation in the 100 years between the mid 18th century and 19th century (1760 to 1830). The economy transitioned from agrarian to industrial, personal incomes rose, cities grew larger, population swelled, and pollution increased considerably.

A confluence of factors created the right conditions for industrialization that elevated England from a regional player in the mid 18th century to the dominant economic power in Europe in a little over a century. Food production increased due to agricultural innovations, which also reduced agricultural wages, prompting large migrations to cities (24). Arguments such as those made by Adam Smith and John Locke for individualism and a market-based economy provided the philosophical justification for capitalism. The invention of the steam engine improved production processes for a number of goods, causing goods to become cheaper and available for mass consumption; it is estimated that there were nearly 2,000 steam engines in use in England by 1800 (22, 24). The falling prices of goods improved living standards and increased savings, which were invested back into companies. At the same time coal, iron, ore, and lumber were in plentiful supply and provided the energy and raw materials for industrial production (22). Labor was also readily available as population increased with declining mortality, particularly child mortality, between the 18th and 19th centuries. Additionally, women joined the factory labor force in large numbers. In comparing conditions in England and China, some have suggested that official strictures in China against employment outside the home limited female labor force participation in nascent factories. Such severe restrictions did not exist in Europe, where women had a history of working outside the home (20).

The industrial revolution was transformative and impacted all aspects of society. There was a decisive and permanent shift toward urbanization. Small, subsistence farms were slowly replaced by larger, industrial farms. GDP, personal income, and wealth all grew; however, some, such as factory owners, benefitted disproportionately. The rise in personal incomes was modest (12% between 1770 and 1820), and productivity gains were also small (0.4% per year between 1770–1830) (27). People became more wage and market dependent, and the labor market grew to include women and children. Environmental pollution also increased, as did greater resource consumption and waste.

The industrial revolution in 1850s Britain is sometimes referred to as the first revolution. The second was the proliferation of electricity as a power source in industrial, commercial, and residential sectors. The third revolution occurred with the widespread use of the computers, information technology, and the internet (25). The fourth industrial revolution will likely combine the digital, biological, and inorganic, physical world and will include the development of technologies based around artificial intelligence (AI), cloud computing, robotics, 3D printing, IOT (internet of things), and blockchain wireless technologies (21). The revolutions are named as such because they radically alter our relationship with time and space. Rising productivity, greater efficiency, and improvements in product/service quality are hallmarks of all industrial revolutions. They also democratize hitherto specialized industries. For example, the second revolution, electrification, allowed us to develop a number of new tools and machines for all aspects of our lives, including our homes. Entertainment also became a consumer product with the advent of electricity. Before the second revolution, entertainment wasn't readily available in homes. After electricity the average consumer was able to listen to music on the radio and watch actors on television. The third revolution took entertainment out of the hands of the professionals. YouTube, Instagram, and TikTok stars and other social media "influencers" command followings and earnings that rival mainstream actors.

Industrial revolutions are job creators and destroyers. The number of factory workers has plummeted in the United States in the last 50 years: Factory workers declined by two thirds between 1960 and 2014 (25). Automation has entered nearly every industry in both developed and developing countries and is likely to accelerate in the coming years. A recent study speculates that nearly half of all jobs in the United States will be replaced by machines in the next 2 decades (26). In other countries, particularly those involved in manufacturing, the replacement rate is particularly high as repetitive jobs are easier to automate. Blockchain technology has the potential to eliminate any "middleman" type roles, such as a real estate title agent, as contracts can be executed without any fear of fraud or default. Additionally, those providing products that can be standardized, such as legal contracts, or jobs that require algorithm-based decision-making, such as customer care or medical diagnostics, are also at risk. ChatGPT, related AI, and their future iterations will impact every industry from coding to creative arts.

A major category of worker that is likely to grow is the **gig worker**, someone who enters a short-term contract with several employers. The gig workers in transportation (e.g., Uber, Lyft) have attracted a lot of attention in recent years due to their inequitable pay structure, but there are many other industries such as food delivery or purchasers that also lend themselves to temporary work. Job fragmentation will increase the gig economy to the detriment of the worker, who will no longer enjoy the same benefits and protections as in a regular job (25). These are just some of the likely impacts of the fourth industrial age. There will undoubtedly be many benefits for consumers and society; however, the negative impacts on worker pay, benefits, job satisfaction, health, and increasing wage and wealth inequality are all factors that need to be proactively addressed.

Technology, Communication, and Globalization

The discovery, by Michael Faraday (1831) that an electromotive force (EMF) can be generated through a variation in the magnetic flux laid the foundation for electrification, the second industrial revolution. Faraday demonstrated that a current could be generated using a changing magnetic field. The Faraday disk was the first demonstration of this principle. Other important developments included the telegraph (1838); the Gramme generator (1870), which was the first commercial generator; arc lamps, which were first commercially demonstrated in 1878; and the incandescent light bulb, commercially developed by both Joseph Swan and Thomas Edison. Electrification involved developments in both stable power generation, which included DC and AC technologies; transmission, which included the electrical grid; electricity storage, which included batteries first developed by Alessandro Volta in 1800; and the devices, such as the light bulb, that capitalized on this new source of power. Individuals like Thomas Edison are often credited with "inventing" the light bulb or other important devices. However, Edison's design was the most successful of several light bulb designs; he no more invented the light bulb than Steve Jobs invented the iPhone. Electrification was the result of the efforts of many different scientists and engineers as well as those who helped make products that used electricity and were commercially viable.

The first power station was built in Surrey, England in 1881. Edison opened the first station in the United States in NYC in 1882. By the turn of the century, there were over 3,000 power stations in the United States, and by 1925, over half of the electricity in the United States was supplied by central power stations. In the short span of 40 years, starting in the 1880s, electric power had become a mainstay of the United States and other industrialized countries.

The development of the electricity grid, which in the United States involved the creation of public utilities, spread the benefits of electricity to the masses.

The first major impact of electrification was on communication with the invention and commercialization of the telegraph. The history of the telegraph involved optical systems in which encoded messages were created in one location and then observed through a telescope at a post no more than six miles away. This system worked well but required labor and weather with good visibility. Samuel Morse, William Cooke, and others helped develop the firm commercial, electricity-based telegraphs in the early 19th century (28). These revolutionized communications, as messages could be relayed instantly over long distances. In the United States, by 1850 each major population center had a telegraph line, and lines were extended to California by 1861, connecting the East and West Coasts. By 1865, Western Union had established a monopoly over the telegraph. Another important 19th-century telegraph-based innovation was the stock ticker, which revolutionized how stocks and commodities were traded by allowing orders to be placed at a considerable distance from the trading floor. This, in effect, created a global marketplace for capital.

The other major communication invention of the 19th century was the telephone, developed by Alexander Graham Bell in 1875 (as with other technologies, there are other claimants to the telephone, but Bell was given the first patent for the technology). The first telephone networks functioned like an intercom system connecting the home and offices of private individuals. It was the development of telephone switchboards, much like the electrical grid, that made the technology available for the masses. The first commercial telephone exchange in the United States began operations in 1878, and by 1915 telephone lines connected the East and West Coasts. It was not until after the First World War, in 1927, that the first transatlantic lines went into operation. Thus, by the end of the 1920s, the communication networks for the global transmission of information were in place. In a hundred years, the world had transitioned to a purely technological method for transmitting information. This revolution, coupled with improvements in regional and transoceanic transportation (e.g., railroad, steam ships, automobile), is seen as the first age of true globalization as people, goods, and information all circulated with a speed hitherto unknown. This increased linkages, interdependencies, and integration within nations and between them. The World Wars would slow this impulse; however, the 20th century was characterized by inventions and innovations that would further bridge the gap between far-flung regions, including perhaps the most important technological developments of the era, computers and the internet.

The World Wars and the Rise of the United States

Empire building, colonization, mercantile capitalism, and technological advances of the first and second industrial revolution created an economically integrated world by the early 20th century (29). Trade accelerated with the development of the steam engine and related technologies. The telegraph and the telephone increased local access to global capital. Some globalization scholars also suggest that the unconstrained migration between 1850 and 1913 created a more integrated labor market than in the modern era when borders and migration have been more curtailed due to more restrictive national migration policies (29).

The impact of this first wave of modern globalization on per-capita income, trade, and migration were remarkable. European per-capita income had increased by more than 70% between 1870 and 1914; in the Americas, per-capita income had doubled (31). Migration out of Europe was also proceeding at a steady pace, with large numbers of British, Irish, Scandinavians, and Italians emigrating to the new world. Capital markets were global as European states financed both infrastructure and business ventures in foreign lands (31). The era of integration, however, came to an abrupt halt with the onset of World War I. The aftermath of the war saw a decrease in interbank cooperation as well as increased restrictions on trade, migration, and commodities exchange (30). World War II decimated Europe but had a profound and transformative impact on the United States. The aftermath of the war saw a steady rise in U.S. economic and military power. The war itself was decisive in ending the Great Depression as it revitalized many industries and helped create new ones, such as those related to atomic energy. The New Deal helped mitigate the impacts of the depression; however, the war supercharged economic growth. The U.S. government took important steps, including substantially increasing military spending and creating new bureaucracies such as the Office for Emergency Management and the War Production Board, to redirect industrial capacities toward military production (32). Even organized labor emerged stronger in the United States, helping counterbalance the power of capital (industry) and government. Unionization reached its peak during the decade after the war ended: In 1954, one third of the workers in the United States belonged to a union (33). Unions helped create the middle class by redistributing corporate earnings to labor through better wages and benefits.

Subsequent to the war, as Europe recovered and the Asian giants (Japan, China, South Korea) had yet to emerge, the United States enjoyed a near monopoly on industrial production. Even during the war, the American economy accelerated at levels since unseen (32). American industrial might was complemented by the reserve currency status of the American dollar, courtesy of the Bretton Woods agreement discussed in Chapter 5. The dollar is still used today for most trade-related and financial transactions worldwide. The importance of the dollar ensures that other nations buy American debt (bonds), which help the U.S. government run consistently large deficits that would have sent most other economies into an inflationary spiral.[1] Migration during the postwar years from rural to urban areas was high as cities provided the markets and capital for transforming these new labor inputs into goods and services. Cities grew, income rose, and the United States, in short, prospered.

Conclusion

The history of trade is as old as the first human civilizations. As urban areas formed, they exchanged goods with each other and formed trade networks that reached great distances. The historical evidence for trade between Mesopotamia and Indus Valley is extensive, though they were separated by 2,500 km (1,600 miles). Globalization scholars resist characterizing the ancient world as globalized, given the low levels of economic integration, but there were certainly hints of integration between far-flung regions as far back as Mesopotamia and certainly by the height of the Roman Empire. Similarly, the economic integration between the East and the West played an important role in European rebirth, or the Renaissance. The rise of the Italian city-states was in no small part due to their trade-exclusive contracts with the Mongol Empire. The *Pax Mongolica* (Mongol Peace) ensured uninterrupted trade between Europe and the East, all the way to China, along the silk routes. The influence of the East on Western economies and culture before the age of discovery and colonization was undeniable. European exploitation of the Americas and colonization of Africa and Asia made Europe economically dominant, a position they had not enjoyed since the Roman heyday. Industrialization and electrification cemented European dominance until the World Wars, when their former colony, the United States, became ascendant. The postwar technological innovations led to the computer revolution,

---

1. Some argue that the dollar is waning with the emergence of a multipolar world and multiple, competing currencies including crypto currencies. This may lead to regional blocs and some level of de-globalization and instability in the short run.

or the third industrial age. Additionally, the latter half of the 20th century included the rise of the Asian economies, starting with Japan, the Asian Tigers (South Korea, Taiwan, Singapore), and then China. The dissolution of the USSR in 1991 was another key event that led to declarations of the "end of history," as some assumed that global capitalism and democracy had won to create a new stable order characterized by economic development, prosperity, and human rights across the planet. Global capitalism would ensure that the world is too integrated and interdependent to engage in self-destructive actions. However, worlds events of the next 30 years would show that the world economy is neither immune from shock, nor is peace inevitable. Additionally, a new threat—COVID-19—would demonstrate the fragility of this ecosystem as the very connections that enable the world economy would result in the rapid spread of a virus with the potential to kill millions.

Chapter 4

Economic Fundamentals of the Modern Global-Capitalist System

Key Terms

Capitalism: An economic system in which scarce resources for goods and services are allocated based on supply and demand in an open market. Capitalism allows for private property ownership and assumes that people are self-interested, profit seeking, and likely to engage in activities that improve their financial standing. Capitalism is often contrasted to centrally planned economies in which government sets quotas for production.

Micro- and Macroeconomics: Economics distinguishes between individual and group decisions regarding production and consumption (micro) and the impact of production and consumption decisions on employment, national income (GDP), inflation, poverty, and other large-scale (macro) economic indicators.

Markets and price: A place, real or virtual, where buyers and sellers come together to exchange goods and services for other goods and services (bartering) or, in the modern era, currency. Efficient markets require buyers and sellers, freedom to enter and leave, competition, accessible knowledge about both buyers and sellers, a reliable monetary system, and stable prices, which emerge from the intersection of demand and supply forces in a market (i.e., prices that are not manipulated but solely the function of demand and supply).

Money: Any item that market participants have agreed on as a "unit of account" so that prices for different goods and services can be paid using whatever serves as money. Anything used as money should be both convenient to use and exchange. In the modern era, money often refers to sovereign (national) currencies like the U.S. dollar. In prison, for example, cigarettes or ramen noodles have been used as money.

Capital and labor: Capital usually refers to money and the means of production (e.g., tools, equipment). Capital is also used to refer to those who control capital (i.e., company or factory owners, the rich). Labor refers to those who work for a wage and are paid for their time by capitalists. In classical Marxism, capitalists (company owners) take the products and services made by labor and turn them into capital (profits), further enriching themselves at the expense of those who created the products and services and are merely paid a wage for their efforts. This distinction is more difficult to make in developed countries as many engaged in labor are also part of the capitalist class through their ownership of company stocks.

Taxes: Money paid to the state to finance the activities of the state. Taxes are collected on income, added to the price of goods and services, and paid on assets like real estate.

Wealth and income: Taxable wages paid for services or labor provided are usually classified as income. Wealth refers to the value of assets such as real estate, stocks, or art that are usually only taxable when sold. Wealth taxes have been implemented in some countries with mixed success as it is difficult to assess wealth outside the context of an actual sale when assets fetch a specific price in a market on a given date.

Inequality: The increasing concentration of income and wealth among fewer individuals and groups. Global capitalism has created increasing standards of living and reduced absolute poverty but also increased both income and wealth inequality. Inequality seems to be directly correlated with social, health, and other problems.

Other key terms will be in **bold** and defined in the text.

Review Questions

- What is capitalism, and how is it different than centrally planned economies?
- What is a market, and what are its assumptions? How do supply and demand function in a market to set prices? How are prices paid in a market?
- What are rent-seeking behaviors, and how do they contribute to inequality?
- What are market distortions? How are distortions related to taxation?
- How are income and wealth different, and what is their relationship to inequality?
- What are demand-side theories, and what do they say about aggregate demand and the role of government?
- What are supply-side theories, and what do they say about money and the role of government?

Introduction

Globalization is primarily an economic phenomenon, though it also involves the exchange of culture (e.g., language, food, ideas, etc.). The previous chapters introduced economic concepts, such as specialization, trade, and economies of scale, essential for understanding globalization as an economic system. This chapter further explores these micro- and macroeconomic foundations of the modern, global capitalist order. There is no one type of capitalism. For example, capitalism in China is under the aegis, or control, of the central government; U.S. capitalism is characterized by large corporations and entrepreneurial activity; Russian capitalism is oligarchic as power and resources are concentrated; the Nordic countries have chosen a more welfare style of capitalism, with the state using heavy taxation to fund a strong social safety net for citizens. Some scholars have suggested a basic dichotomy between liberal market economies in which governments don't interfere much in markets and coordinated market economies, which subordinate firms more heavily to nonmarket forces such as regulations and collective bargaining agreements (16). There are other examples as well; however, no system is only defined by one type of capitalism. Irrespective of their differences, the global capitalist system has certain key features such as reliance on markets to set prices, use of sovereign currencies to pay for goods and services, a central banking system that regulates the money supply to speed up or slow down the economy, and the use of taxation to promote or reduce certain behaviors (i.e., to induce a market reaction).

The Economy

James Carville, the chief political strategist for U.S. President Bill Clinton, coined the catchphrase "It's the economy, stupid" to keep campaign workers focused on one of the three central messages of the Clinton presidential campaign. Not a day passes without hearing some reference to the economy in conversation or the media. However, what is the economy? If pressed, most would employ concepts such as trade, production, price, money, goods and services, income, and wealth to describe their understanding of the economy. More sophisticated explanations may invoke monetary and fiscal policy. These and other concepts are certainly part of the **economy**, but they don't capture its *sine qua non*, or most essential feature: the utilization of scarce resources to produce goods and services.

Resources can be anything and typically include land, labor, capital, entrepreneurs, and technology. The optimal use of resources to create and sustain economic growth is the central question facing both public and private sectors throughout the world. In some countries the public sector, or government, plays a significant our outsized role in economic planning, whereas others opt for a more **laissez-faire,** or noninterventionist, approach. Economic growth and level of economic development are inextricably linked to the standard of living in a region. The **standard of living** is linked to the variety and number of goods and services available for purchase in a region. Usually, the more goods and services a household can buy given prevailing income levels, the higher the standard of living in that region. Another important and related concept is poverty. Those below the **poverty line** do not have the economic capacity to consistently procure a minimum level of nutrition and other basic necessities such as housing and health care. Poverty can also refer to having access to sufficient necessities but not enough for discretionary spending on self-improvement, recreation, or consumer goods. Another measure is the **Human Development Index (HDI)**, which combines an economic indicator (GDP per capita) with health (life expectancy) and education (literacy) indicators. The HDI is a better measure of human well-being as it looks at access to important nonmonetary sources of quality of life. Some countries, such as Sri Lanka or Gabon ,score high on the HDI despite being relatively poorer in terms of GDP or per-capita income; also, income distribution is unequal in both countries. The terms **developed and developing, or first and third world,** have been used in the past to differentiate between industrialized economies with a high standard of living and those that are less developed or with a lower standard of living. These terms can be problematic as they do not capture intra-regional (or within-country) variation in income, poverty, and related issues (e.g., developed or first-world countries can have significant income inequalities, or some groups within the countries suffer disproportionately from health or other burdens); additionally, the terms carry the baggage of colonialism, imply some sort of hierarchy between countries, and paint entire countries with a single, pejorative (negative) brush. These issues notwithstanding, the terms are still used in both academic and policy literature and colloquially to differentiate between high-income and lower income countries. According to the World Bank, in 2021, countries with a **per–capita income** (GDP divided by population) of less than $1,046 were low-income, those between $1,046 and $4,095 were lower middle income, those between $4,096 and $12,695 were upper middle income, and those exceeding $12,695 were considered high income (19). These ranges are useful for classification; however, specific attributes are more useful from a policy perspective to address specific indicators of well-being or human development, such as household income, poverty, wealth and income distribution, health care access, morbidity/mortality, or educational attainment.

This chapter provides an overview of concepts essential for understanding markets, income and wealth, and the role of the government and central banking in managing the economy. In this context, we will review both **micro-** and **macroeconomic** concepts such as markets, price, money supply, and taxation. Subsequent chapters will explore the financial and banking sectors, including a detailed discussion of the Federal Reserve system, and world trade, including a discussion of the IMF, World Bank, and World Trade Organization. This chapter will also briefly discusses inequality within the context of wealth effects. Chapter 6 will provide a more detailed discussion of poverty, inequality, and human development.

Capitalism

Globalization or **global capitalism** is a macroeconomic phenomenon. **Macroeconomics** is the study of large economic systems, including national economies and how they interact to create the world economic system. Macroeconomic theories are typically concerned with economic growth (or lack thereof), employment, inflation (and deflation), and trade. Though this section is primarily concerned with macroeconomic questions, it is important to understand that macroeconomic trends are the result of the multitude of decisions made at the microeconomic level. These decisions interact with each other to structure regional and national economies, which collectively shape the global economic system.

Capitalism lies at the heart of any discussion about globalization. **Capitalism** is an economic system that privileges individual decision-making, private property ownership, and the market. In a capitalist system, individuals and groups can own land and other tangible goods such as art or jewelry, as well as more abstract or intangible items such as copyrights, shares in a company, or, today, nonfungible tokens (NFTs). This perspective can be contrasted to, for example, a **usufructuary** or communal concept of land ownership in which land can be used but not owned individually. Capitalism also assumes that people are self-interested and profit seeking and that they freely engage in activities that improve their financial standing. **Private property ownership** under capitalism allows for the private ownership of assets such as land and technology required for making finished products as well as intellectual property. Property ownership under capitalism includes the right to use, exclude, transfer, and destroy. This is not the only conception of ownership but the one most closely aligned with global capitalism.

Labor, in modern capitalism, is exchanged for wages, usually some type of currency. Capitalism envisions a small role for government, one that is limited primarily to protection of civil rights, the provision of defense and other services that the market will not provide, protection of common pool resources such as water, and enforcement of contracts. For all other goods and services, there are markets that can be entered freely by both buyers and sellers. Demand and supply respond to each other by creating a **market clearing price**, or the price at which demand equals supply for a particular good or service. Early economists such as Adam Smith, who coined the phrase "**the invisible hand of the market,**" believed that markets were ultimately efficient and did not require intervention and would ultimately create full employment. The Great Depression waylaid these ideas and paved the way for **Keynesian** or demand-side economics, which argued for significant government intervention to stimulate aggregate demand when the economy had stalled. Later, during the hyperinflation of the 70s, Keynesian ideas were replaced with monetarist thinking (i.e., Milton Friedman, Hayek), which emphasized control of the money supply. We will return to Keynesian and monetarist ideas later in this chapter.

Criticisms of Capitalism

Important criticisms of capitalism come from **Marxist theories**. One argument is that capitalism alienates labor (or people) from the products of their labor; capitalism allows the owners of the means of production (i.e., capitalists) to appropriate products made by others and turns them into **capital**, or money. Capitalists take surplus value, or profits made by selling products made by others, and use them for their own ends; workers don't have any control over products they have made. This leads to growing inequality between labor and the capitalists and, in the long run, growing malaise among the general population characterized by powerlessness and loneliness (4). Other research indicates that BMI, smoking, and excessive alcohol consumption are inversely related in the United States: Poorer households are more stressed, more likely to drink and abuse alcohol, more likely to smoke, and more likely to be overweight and obese. These factors increase the risk of morbidity (or sickness) from a variety of ailments such as cardiovascular disease, diabetes, and cancers and lead to lower life expectancy (9, 10). In the United States, states with the most income disparities also have the lowest life expectancies (10).

Economic data from throughout the world confirm growing income and **wealth inequality** across different countries. Many explanations have been proffered for inequality, including a recent analysis by French economist Thomas Piketty, which argues that the growth rate of capital (i.e., investments) exceeds economic growth (i.e., GDP) (3). The former (capital growth rate) benefits capitalists, or the wealthy who invest their profits for a higher rate of return in stocks, bonds, and the like. The latter (economy growth rate as measured by the increase in GDP) impacts wages paid to labor, which grow much more slowly than capital. In other words, investments grow faster than wages, which benefits those with exposure to investments. The explosive growth in the stock market after the 2007–2008 recession and much slower GDP and wage growth in the United States illustrate this phenomenon. Given the effects of **compound growth**, the wealthy grow wealthier more quickly than others, creating the inequalities seen across the planet. This argument is not without its detractors. However, irrespective of the reasons for growing inequality, the negative impacts are undeniable as an increasingly smaller group of people controls larger portions of the world economy. The trend towards private equity will likely exacerbate these inequalities (i.e. companies becoming privately held and outside the scrutiny of regulators overseeing publicly traded companies). The growing chasm between the wealthy and others is perpetuated intergenerationally as wealth can provide education, social capital, access to networks, and other opportunities to those from wealthy families, allowing them to earn more wealth than those without the same advantages.

The U.S. Federal Reserve's Survey of Consumer Finances (CSF) provides insight into the characteristics of those on the receiving end of wealth transfer. Not surprisingly, those inheriting wealth and gifts tend to be more educated, higher earning, and wealthier than the average American (5). Wealth thus compounds inequality over time, as those who receive wealth transfers from previous generations have greater access to wealth-generating inputs such as education and inherit wealth, which can be invested to keep increasing its size. Wealth also allows subsequent generations to take greater entrepreneurial risk as wealth can mitigate the impact of failure.

Between 1995 and 2016, there were roughly 1.7 million transfers of wealth annually in the United States. Of these, the vast majority were far below $50,000. Those receiving over $1 million in inheritance only accounted for roughly 2% of the total number of transfers (6). However, when looking at the share of the total inheritance pie, those inheriting $1 million or more accounted for 40% of the total inheritance. In other words, those inheriting substantial wealth are numerically small, except their wealth is significantly greater than others receiving smaller amounts. This trend points to a further concentration of wealth over time in the United States fueled, at least partly, by inheritance and gifts (6).

The concentration of wealth and its passing to subsequent generations often creates "**rent–seeking behavior**." A rent in the classical economic sense is any wage or price that exceeds the market price for labor or a good. An example is the exorbitant salary paid to an actor who would be willing to work for much less. The excess salary is economic rent. Economy rent also refers to any actions taken by the wealthy to protect their wealth and increase their wealth without increasing economic productivity. There is ample evidence for rent-seeking behavior in many different areas of the economy. The investment and financial sectors, for example, allocate capital to unproductive areas while charging fees for doing so; real estate rents have increased significantly in many countries, exacerbating a housing affordability crisis; pharmaceutical companies charge usurious prices for certain medicines that provide very little benefit in terms of improved morbidity or mortality. In short, rent-seeking behavior increasingly privileges the wealthy at the cost of economic innovation and productivity.

Rent–seeking behavior includes laws and regulations that favor the wealthy, which also increases market distortions and inefficiencies, leading to suboptimal market prices. For example, California's Proposition 13, passed in 1978, limits property taxes to 1% of the assessed home value, which cannot be adjusted upward by more than 2% annually irrespective of actual price appreciation. Proposition 13 benefits existing homeowners as the promise of low property taxes that are both capped and not linked to actual home value reduces the total payments associated with the home, leading to greater price appreciation. As homes increase in value, so does homeowner wealth. This wealth can create disparities in consumption as homeowners can tap into their home equity through a line of credit or "cash out" mortgage refinance without incurring any type of penalty or **capital gains tax** (tax levied on profits).

Recent work suggests that information rents, or those rents deriving from the "commodification of information and knowledge," improve labor productivity and increase economic growth, but they do not maximize value-added growth (i.e., the impact on the economy) and benefit only high earners (7). Wealth concentration thus likely acts as a drag on overall economic growth, which reduces wage growth, impacting the nonwealthy, wage-dependent households disproportionately. International Monetary Fund (IMF) studies also show that wealth inequalities, as measured by the GINI coefficient, significantly impede economic growth and increase the risk for economic, financial, and political instability. The GINI coefficient is a commonly used measure of income distribution. GINI values vary between 0 and 1; the coefficient is zero when everyone has the same income and 1 when one person controls all income. The closer a region's value to 1, the more unequal its income distribution. According to the IMF, global inequality is high and measures somewhere between 0.55 and 0.7. The U.S. GINI of 0.49 indicates a society that is somewhat more unequal than its developed counterparts.

Inequality can be conceptualized in terms of disparities in income and wealth or unequal opportunities. The two are linked as lack of opportunities can impact individual capacity for increasing income and wealth. A healthy market is predicated on a "level playing field" so that distortions are minimal and individuals or groups can enter and leave freely. If structural issues, (i.e., the institutions and policies in a region) privilege one group over another or create barriers to market entry, in the long run inequalities will worsen along with the productive capacity of the region. Inequality begets or creates more inequality and stifles economic productivity and growth. Many have argued that greater wealth and income at top tiers is acceptable as spending from the top eventually **trickles down** and positively impacts the rest of society (this is consistent with supply-side economic theory). The evidence for **trickle–down the-**

ories is sparse at best. A recent study of 18 OECD countries suggests that trickle-down policies further enriched the top 1%, without any positive impact on economic growth or unemployment (17). It is critical to address inequality, poverty, and related issues to create a more sustainable world. These concepts will be discussed in greater detail in Chapter 7 when we look at economic development and poverty policy.

Micro- and Macroeconomics

The preceding discussion on capitalism and inequality contained several important economic ideas, such as the market, rent seeking, and inequality. Economics often distinguishes between two broad domains, the micro and the macro; academic economists often identify with one or the other. The distinction, however, is more conceptual than practical as macroeconomic phenomena are a function of the many decisions and actions taken at the micro level. Adam Smith and David Hume, along with others, provided many of economics' foundational ideas. Smith, in the Wealth of Nations, published coincidentally in 1776, the same year as the American Declaration of Independence, argued that because individuals are inherently self-interested, they flourish in a **largely unregulated free-market system** as regulatory actions create distortions that result in shortages or surpluses. **Mercantilist** thought dominated economic discourse during Smith's time. Its proponents argued that since total wealth was finite or limited, policies should encourage exports and discourage imports to increase inflows of capital (or gold) into the country. Pursuant to this thinking, many countries imposed heavy tariffs or taxes on imports. Smith and Hume correctly recognized that such protectionism would eventually reduce trade and cause **inflation** as capital (or gold) inflows into a country would increase prices for goods and services in the country. They both recognized the importance of private property, though Hume did not think it was "natural right," like John Locke, but instead argued that property rights were necessary in a resource-constrained environment. Smith reasoned that **specialization** and **division of labor** were necessary for increasing productivity as workers who focused on specific tasks would improve at the task over time, increasing their efficiency through repetition or practice, and find new ways, including investing in technology, to do so.

Smith also proposed a **theory of value** based on labor, specifically that the value of goods or services is based on the amount of labor required for production. Things that required more labor would, thus, cost more. Over time competition in a free market would lead to innovation, reducing the labor hours required and drive down prices. This would lead to a greater variety of cheaper products, a win for consumers. At the national level, the ability to produce certain goods more efficiently was essential for trade as countries could specialize in products and services most suited to them and then engage in trade. Later, David Ricardo would expand this analysis to include indirect costs, or **opportunity cost**, in his discussion of **comparative advantage**, a key tenet of globalization.

Hume's analysis of money in Essays, Moral and Political (1754) presaged modern monetary theories. He reasoned that the actual supply of money was not important as prices would rise or fall in response to changes in supply (13). If more money was injected into an economy, the result would be an increase in prices or inflation. Expanding on this idea in his "Specie-Flow" thesis, Hume argued that a local increase in money supply will cause inflation, leading to increasing consumption of cheaper priced imports and unbalanced trade (i.e., when the value of imports exceeds exports). This would eventually cause an outflow of money as imports increase, reducing the money supply at home and, eventually, prices, which would then make exports favorable again (13). Smith and Hume provide many of the ideas that would form the basis of microeconomics and macroeconomics. Others such as David Ricardo and

John Maynard Keynes, who will be discussed in the macroeconomics section, would add to the growing literature on economics, providing the firmament for the modern, global capitalist system that we take as natural.

Microeconomic Concepts

Microeconomics, as the name suggests, is concerned with decisions made at the individual or group level (e.g., households or firms) regarding production and consumption. **Producers** are those who make things for sale and who sell services and are engaged in production. Everyone purchasing their goods and services are **consumers**. People and groups can be producers in one context and consumers in another. Microeconomics examines how individuals and groups structure decisions in response to prices and incentives (any action taken to increase consumption). Microeconomic theories generally assume that individuals and groups act rationally to maximize their utility from the consumption of goods and services. **Utility** is sometimes used interchangeably with happiness or satisfaction derived from consumption of goods and services, though this is not completely accurate. In other words, people buy things—goods and services—that they both need and desire. **Need**, however, is difficult to define as people can choose to consume higher cost versions of basic necessities such as food, water, and shelter even when lower cost versions will suffice. All decisions entail a **direct cost**, or cost associated with buying a good or service, as well as an **opportunity cost**, the indirect cost associated with forgoing, or not consuming, other goods and services. Put another way, the direct cost is the toll exacted by the road taken, and opportunity cost includes the benefits (e.g., better view or less traffic) associated with the other roads not taken. This is an important concept to which we will later return as it is essential for understanding comparative advantage, trade, and globalization.

While individuals consume various combinations of goods and services to maximize their utility, producers of these goods and services seek to minimize their **input costs**, or costs associated with production, to maximize their **profit**, the difference between production cost and market price for the good or service. Supply and demand for goods and services determine the market clearing price. A **market** is a place, real or virtual, where buyers and sellers come together to engage in trade. A well-functioning market has a few important requirements: a large number of buyers and sellers; freedom from coercion; perfect competition and knowledge; sound monetary system; and, perhaps most importantly, one price. Perfect competition, though it doesn't exist in practice, is perhaps a market's most important attribute. In a market with **perfect competition**, there is no cost associated with buyers and sellers locating each other and engaging in trade; additionally, neither side has the ability to manipulate prices as the **market clearing price** emerges from the intersection of demand and supply, specifically when demand and supply are in equilibrium or when the quantity of a good or service supplied is equivalent to the quantity demanded.

Price is the amount of something a good or service is exchanged for in a given market. Price is often paid using **money**, usually a sovereign currency, though this is not always necessary. **Money** can be anything that market participants have agreed on as a **unit of account** so that prices for different goods and services can be paid using whatever serves as money. Anything used as money should be both convenient to use and exchange. It is a lot easier to carry and exchange paper money than most other items (e.g., gold) that have historically served as money. Money should also retain its value so that it can be used in the future to buy things. Later in the chapter, we will discuss inflation, which erodes the value of money. Money should be fungible so that the thing used as money can be replaced by another similar thing without any loss of use as a medium of exchange and value. Precious metals such as gold and silver have been used as money, as have more unusual items such as cigarettes, which were used as currency in parts of Europe following WWII; cigarettes have also been used as currency in prison, though in the last few years they have been supplanted by ramen and other cheap food items as privatization of prisons has shifted the burden of adequate nutrition to prisoners and their families (1).

In the last few years, crypto currencies, led by Bitcoin and Ethereum, have become more mainstream. Some have argued that cryptos represent a decisive break from fiat currencies, or money issued by sovereign nations, and the old economic order. Additionally, proponents argue that cryptos are a better store of value as they cannot be manipulated through the traditional tools of monetary policy that include printing more money, which often devalue currency by creating inflation. In Chapter 5, we will look at the various monetary policy tools typically used by governments and their relationship to **inflation**, or the tendency of price of goods and services to rise with time. Others have forcefully rejected the idea of cryptos as a currency or a store of value, citing the difficulty of using cryptos as payment, volatility in price, and technology dependence (2). Others cite the distributed nature of crypto currency block chains, which are located on various computers in the crypto network, as a source of their strength as no nation or single entity can control or manipulate them.

If a good is in high demand but supply is low, producers can charge a higher price for the good until supply increases as a result of more goods being made by the existing producers or newer producers coming into the marketplace. Supply is a function of several factors: availability, cost, and productivity of resources, technology, and labor; profitability of alternative goods; environmental factors (these are extremely important for food-related goods); and competition. Similarly, demand can vary because of changes in consumer preferences, the cost and quality of alternative goods, and changes in consumer income and wealth. Advertising or marketing also plays an important role in consumer demand. In 2021, over \$250 billion was spent on advertising in the United States, the world's largest consumer market (18).

The relationship between price and demand is sometimes referred to as the **price elasticity** of the product. A highly elastic product is one in which a small change in the price of the product causes a significant change in demand for the product. Conversely, a highly inelastic product is one in which even large changes in price don't impact demand significantly. Paid subscriptions for social media are highly elastic. If Facebook, for example, were to introduce a monthly surcharge for its services, it's likely that users would migrate *en masse* to other, free platforms. In contrast, essentials such as energy, water, and food are more inelastic, at least in the short term. Changes in energy, water, and food prices would not initially alter demand significantly, though in the long run consumers may reduce consumption or seek alternatives.

As markets drift away from these requirements, **distortions** arise that impact the market price for goods and services. Most distortions are the result of some type of government intervention typically undertaken to protect market stakeholders, which can also include the environment, or enhance their well-being. Government policy often tries to strike a balance between regulation, which introduces distortions, and noninterference in markets. Government intervention is sometimes necessary to prevent **monopolies**, or market dominance by one or a few companies. Monopolies can stifle competition and lead to prices well in excess of what would occur in a normally functioning market. In some cases, monopolies arise because of barriers to market entry, which can include high input costs (e.g., power or telephone lines, satellites). In some cases, it makes sense to grant a monopoly to one producer given the technical or other hurdles (e.g., legal) involved with production and provision (e.g., water or energy supply) and then regulate the producer to prevent usurious, or excessive, charges. Companies providing electricity, for example, in the United States are heavily regulated as public utilities because they are **natural monopolies** as they are often the only or one of a few providers of electricity in an area. Governments also grant temporary monopolies, sometime called **copyrights**, to certain types of products such as music, literature, or software as they are easy to copy and distribute. Copyright allows the developers of these products to profit from their enterprise, which helps incentivize further production and innovation in these markets. Without such protection, firms such as Microsoft, one of the largest and most profitable companies in the world, couldn't exist and thrive (though some would argue, Microsoft has a monopoly on operating systems). Government regulation is also necessary to prevent or mitigate **externalities** or impacts on those not part of the market transaction. An example is plastic pollution and its impact on ocean ecosystems (ocean flora and fauna are not part of markets for various plastic goods).

Income and Wealth

It's important to differentiate between **income and wealth** as the two are often used interchangeably. Income refers to money earned by any entity, individual, group, or company in a particular time period. Wealth refers to the value of assets held by individuals, groups, or companies. Wealth typically includes real estate, which can include undeveloped or developed land, and equities (stocks), but can include anything that has an **exchange value**, which colloquially refers to the price of the asset in a market. The exchange value is often contrasted to **use value**, which refers to the value derived from the use of a particular thing. Ironically, sometimes things with a high use value, such as water, have low exchange values (or price in a market), and vice versa (e.g., diamonds). According to David Ricardo, who is also credited for developing the concept of comparative advantage, an essential component of globalization, the exchange value derives from the relative scarcity of an item and the labor embodied in its creation. The value of art created by a master artist, especially one that is deceased, is based on its scarcity. The value of a large, custom home is based on the quantity and quality of labor required for its construction as well as the relative scarcity of the land on which the home is built.

Things like precious metals, jewelry, art, or NFTs (nonfungible tokens) can all function as assets as long as there is a market for them. Investors will often diversify their assets by purchasing assets such as art or gold as a hedge against fluctuations in the prices of more traditional assets such as equities (stocks) and real estate. It's possible for individuals and households to have very little annual income while having substantial assets that make them wealthy. A common example in the United States is an elderly couple with only a meager Social Security income who lives in a million-dollar home without a mortgage. Their wealth would include the home's equity or the value of the home minus the remaining mortgage (which in this case is zero). These households are sometimes referred to as "house rich and cash poor."

Individuals typically earn income from exchanging their labor for a wage (i.e., income from work). In the United States, income earned by individuals is usually reported yearly on a W2 or 1099 tax form. The W2 form shows gross income, or the total income paid for labor by the company issuing the W2, and deductions, which include federal and state payroll taxes as well as Medicare (health) and Social Security (government pension) contributions by the individual and employer. The form also contains deductions made for state taxes and other parts of the social safety net (e.g., unemployment insurance). A 1099 form is issued by a company for services usually performed by contract labor such as a vendor or consultant who is not part of the company's regular employee pool. There are a few different kinds of 1099 forms, with the most common being the 1099-NEC or nonemployee compensation (previously called the 1099-MISC). Payroll taxes are not deducted from income reported on a 1099 form. The individual receiving the form is liable for paying payroll taxes quarterly or annually. A common criticism of 1099 income is that those receiving income as a contractor or vendor are responsible for contributing both employee and employer share of the total Medicare and Social Security contributions. Some employers use the 1099 option to reduce employee-related costs as they don't have to provide vendors or subcontractors with the same benefits (e.g., sick leave, vacation) as regular employees. Many states impose substantial fines and other penalties for classifying regular employees as contract or 1099 workers.

Macroeconomic Concepts

Macroeconomics is concerned with the impact of consumption and production on employment, income, wealth, savings, and price levels at the macro, or large, level, usually the national or international level. At a minimum, this involves studying the economies of smaller regions within countries. Economists use the term **GDP**, or gross domestic product, to measure the economic output of a country. GDP is the total value of all goods and services produced in the country in a certain time period.

There are other ways of measuring national income (e.g., gross national income), but GDP is the standard used by most countries. The economic output of other countries is converted from the local currency to dollars at current exchange rates (i.e., the amount of currency that can be purchased for one dollar). In some cases, an adjustment is made for variations in the cost of goods and services locally. Simply put, a dollar in a country with a lower cost of living, usually a poorer country, can buy a lot more than a dollar in a comparatively higher income country with a higher cost of living. This adjustment, called **PPP** (purchasing power parity), is based on the difference in cost of an equivalent bundle of goods (e.g., haircuts, food) in both countries. For poorer countries with lower costs of living, the PPP GDP is often much larger than the **nominal GDP**, which is calculated using exchange rates. The largest economy in the world using nominal GDP is the United States; however, China takes the number one spot once PPP adjustments are made. Similarly, India's economy jumps to third place, overtaking Japan once PPP adjustments are made.

Another important statistic is **per–capita income**, which is calculated by dividing the GDP by a country's population. This is a crude measure as it does not provide any insight into how income and wealth are distributed. It's possible to have a high per-capita income and high levels of economic inequality, as in the United States. Another important measure of income is household income, which is estimated by combining the incomes of all members of the household. Households can be of any size, including single-person households. Household income and the percentage of total household income controlled by different household quintiles provides an important measure of the income distribution in the United States. (Chapter 6 explores households and other demographic concepts and data in greater detail,) The federal poverty rate in the United States is also estimated by looking at household incomes. There are also other measures of well-being and human development (e.g., HDI) not explicitly tied to income. Chapter 7 provides a more detailed discussion of poverty, inequality, and human development in the United States and internationally.

Macroeconomic actions refer to polices and actions taken by policymakers to steer the national economy. The relationship between macroeconomic variables is often difficult to understand and conceptualize. Nonetheless, those in macroeconomics formulate theories and **empirically** evaluate their theories by collecting data. **Econometricians**, researchers who test theories using statistics, specify the relationships between a set of variables in some type of regression model, in which some variables are independent or causal, and others are dependent or influenced by the independent variables. A key feature of regression models is their ability to determine the relative, causal, impact of different independent variables on a dependent variable of interest. For example, researchers can evaluate the relative impact of independent variables, such as education, race, and gender, on income. Econometric modeling using large data sets, such as those collected by the Bureau of Labor Statistics, the General Social Survey (GSS), or the U.S. Census, provides valuable insight into the explanatory power of macroeconomic theories. Regression models can help us understand which variables to focus on through policy when trying to address important social problems like unemployment, consumer spending, or entrenched poverty.

Macroeconomics is also concerned with understanding large-scale economic cycles, which can include periods of economic downturn such as recessions and depressions. In the United States, the **National Bureau of Economic Research (NBER)** is tasked with identifying **recessions** (we have had 34 so far) or two consecutive quarters (two, 3-month periods) of negative GDP growth. **Depressions** are more severe and can last much longer: A 10% contraction in the real GDP is sometimes used as the benchmark for a depression.

The stock market crash of 1929 and the ensuing depression (the only one experienced by the United States to date) were poorly understood using existing economic ideas. Existing public policy was thus inadequate to stimulate the economy. John Maynard Keynes, widely heralded as the father of macroeconomics, in his landmark books, A Treatise of Money, and its follow-up, The General Theory of Employment, Interest and Money, argued that government was necessary to stimulate **aggregate demand**, the total spending by households, businesses, and government. This upturned decades of prior thinking that held that unfettered markets would eventually create full employment. Keynes reasoned that the production of goods and services depends on consumption, investment, and exports and that government could and should impact all three through its own expenditures, policies, and programs. Keynes proposed that falling consumer confidence decreases household spending. In response, businesses reduce investments in production, including labor, leading to reductions in output but not price. By shedding labor, usually the biggest piece of production costs, businesses reduce size while maintaining prices. Reductions in labor lead to higher unemployment, which further reduces consumer confidence and perpetuates the contractionary cycle. The market, Keynes argued, becomes stuck on the ground floor. Active government intervention in the form of increased expenditure and stimulatory fiscal policy (e.g., tax-based incentives) is needed to jumpstart the economy. Government can increase spending, which also has a **multiplier** effect (i.e., every dollar spent generates more than a dollar of economic effect) to purchase goods and services, often through large infrastructure projects; create tax-based incentives for households and businesses to spend; and inject capital (i.e., bailouts) to stabilize certain sectors of the economy. Monetary policy, which impacts the actual money supply, can be used as a complement to stimulate growth. The Federal Reserve or Central Bank usually sets monetary policy. In the United States, the mission of the Federal Reserve, or Fed, is to "promote effectively the goals of maximum employment, stable prices, and moderate long-term interest rates." (14). The Federal Open Market Committee (FOMC) is thus tasked with curbing excessive (more than 2% annually) **inflation**, which is the tendency for the price of goods and services to increase with time, and maintaining low **unemployment**, or the proportion of workers who are seeking work but cannot find it. In the United States, the **Bureau of Labor Statistics (BLS)** measures unemployment monthly through a household survey. In some cases, those who were previously looking for work stop actively seeking work due to fatigue; low wages in the areas they are looking; becoming eligible for retirement or similar benefits; and sickness. Another important issue is **underemployment** as some take part-time, hourly, contract or lower wage work. Measuring unemployment can be tricky, but it is an important **metric** or measure of the economy's health. Typically, economic growth reduces unemployment. Economists have sought to quantify this relationship by looking at unemployment levels when the economy is expanding or contracting. However, irrespective of growth, there always seems to be some level of unemployment, raising the question of the lowest possible unemployment rate in a certain economy. This is an important question for policymakers as they seek to adjust fiscal and monetary strategies to achieve low unemployment.

The Federal Reserve uses monetary policy to reduce unemployment and increase economic activity, but also to curb inflation, the tendency of prices to rise with time. The Federal Reserve impacts the money supply through one of three principal avenues: the federal funds rate, the reserve ratio, and open market transactions to buy and sell assets (securities). This process is also sometimes called QE, or quantitative easing. The various roles of the Federal Reserve will be discussed in greater detail in the next chapter ,along with banking and the financial sector.

The monetarists, unlike the Keynesians, argue for either no response to an economic downturn, as these are viewed as normal and cyclical, or conservative actions to minimally impact the money supply. Many strict monetarists advocate for pegging the growth of the money supply to the **GDP** as a way of limiting the growth of the money supply and, ultimately, inflation. An increase in money in the system without any changes in the supply of goods and services means more money will be chasing the same amount of goods and services, hence demand will exceed supply, causing inflation. The **quantity theory of money**, the basis for monetarist policy, postulates that the money supply, multiplied by how often money changes hand, or its **velocity**, determines **nominal** expenditures, the price of goods and services multiplied by the number of goods and services. Velocity is viewed as relatively stable by monetarists, hence their focus on restraining the growth of the money supply. In the 1970s, many countries experienced slow GDP growth and inflation. This was termed **stagflation** colloquially. The remedy did not come from fiscal policy or Keynesian approaches but instead from a tightening of the money supply. Tighter monetary policy that restricted the availability of money worked, and runaway inflation was reduced, though the U.S. economy experienced a significant recession, which ushered in the Reagan era characterized by deregulation and reductions in "big" government.

The **money supply**, as you may have inferred, is simply the total amount of money in the economy, which includes cash, coins, bank deposits, and short-term investments (e.g., a 6-month CD or cash in an investment money market account) that can be converted into money quickly (15). However, estimating the money supply is not straightforward. Money supply can be measured by assessing the total currency in circulation and deposits by banks at the Federal Reserve (monetary base), the total cash circulating among the public and in deposits in banks by the public (M1), and M1 along with time deposits or CDs and money market investment accounts (M2) (15). A related term often used in finance is **liquidity,** or the ease with which noncash assets can be converted into cash without a significant erosion of value. Assets such as stocks and real estate have market values on any given day. However, the sale of these assets can sometimes become complicated, as with real estate, and can trigger significant tax liabilities, reducing their cash value. Additionally, when assets are sold during times of economic duress (e.g., recession), prices are often heavily discounted—sometimes called a fire sale—as institutions need to raise cash to remain solvent. This creates significant distortions in the market for the assets being liquidated. If the event is catastrophic, like the 2007–2008 housing-led recession, government intervention is required to contain the fallout and prevent further damage to other sectors of the economy.

Fiscal policy typically involves the use of taxes to promote, reduce, or prevent certain actions. Taxation almost always introduces significant market distortions. Tax-based interventions in the United States include the ability to deduct mortgage interest, which incentivizes home buying, tax credits for purchasing solar or electric vehicles, or the heavy taxation of tobacco to reduce consumption. Thus, taxation or fees and surcharges, which act like taxes on income, goods, and services, significantly impact both supply and demand for goods and services. Though taxes are the primary way government intercedes in markets, other mechanisms, such as direct grants and subsidies, or regulatory actions, such as prohibitions are also used.

A tax credit is a "dollar-for-dollar" reduction in tax liability, whereas a deduction reduces gross income, or income subject to taxation, by the amount of the deduction. Generally, credits are preferable to deductions. As shown in the example that follows, a credit of $10,000 on a gross income of $100,000 taxed at 20% would result in a savings of $10,000, whereas a deduction in the amount of $10,000 on the same income would result in a savings of $2,000.

Tax Credit Example

Taxes owed: $100,000 income × 20% tax rate = $20,000 owed in taxes

Reduction in taxes: $20,000 owed – $10,000 credit = $10,000 owed in taxes after credit

Total savings: $20,000 – $10,000 = $10,0000

Tax Deduction Example

Taxes owed: $100,000 gross income × 20% tax rate = $20,000 owed in taxes

New taxable income: $100,000 – $10,000 deduction = $90,000 taxable income after deduction

Taxes owed after deduction: $90,000 × 20% tax rate = $18,000 owed in taxes

Total savings: $20,000 – $18,000 = $2000

In summary, government, through **fiscal and monetary policy**, can play a decisive role in stimulating economic growth and employment and also slow down the economy when it becomes overheated and inflationary. **Fiscal** action usually involves lowering or raising taxes, and **monetary** action usually involves lowering or raising interest rates by the Federal Reserve or central bank. Government action is also sometimes needed to safeguard the economy and restore normal market functioning by "bailing out" certain sectors of the economy and providing appropriate fiscal and monetary interventions. This was the case during the Great Depression (1929–39) and the 2007–2008 great recession. In one sense, the central debate between Republicans and Democrats in the United States is macroeconomic, as Republicans traditionally champion less government intervention and Democrats generally favor a more activist government (these are simplifications but capture, at least, each party's explicitly stated stance on government involvement in markets).

Conclusion

Most countries use some type of a market-based economy. However, there is no one type of free market economy or capitalism. The role of the government varies, with some countries favoring minimal government involvement in markets and others favoring state-directed capitalism or heavier regulation and more taxation for government spending on the public safety net, externalities engendered by capitalism (e.g., environmental impacts), infrastructure, defense, and other national priorities. The sheer variety of "capitalisms" suggests that market-based thinking has triumphed. However, there are significant problems within successful economies and between nations. Economic inequality is widening, though extreme poverty is declining. Paradoxically, the standard of living, as measured by health (e.g., life expectancy), social (e.g., educational attainment), and other indicators (e.g., access to technology), of those in the lower income ranges has increased, though real wages have grown slowly or stagnated. The destructive impacts on the environment are undeniable as resource consumption increases along with pollution and climate change impacts. So where does this leave us? Can these problems be addressed within the framework of global capitalism, or is a paradigm shift necessary? In the next chapter, we will explore the institutions of global capitalism tasked with maintaining economic stability and growth. In subsequent chapters, we will look at sustainability through the lens of economic poverty and inequality, environmental degradation and climate change, and other social impacts, including public health. We will end our journey by examining what we can do as individuals to create a more sustainable world.

Chapter 5

The Institutions of World Trade (IMF, World Bank, and WTO) and the Modern Financial System (Banks, Stocks, and Regulation)

Key Terms

Comparative advantage: Increased efficiency in the production of certain goods and services due to advantages in labor, technology, resources, or other favorable local conditions. Countries should specialize in those goods and services for which they have a comparative advantage and engage in trade for other goods and services for which others have a comparative advantage. This process results in an increase in the quantity and quality of goods produced worldwide.

International Monetary Fund (IMF): Created in the aftermath of WWII to stabilize trade and the world economy by providing loans and economic advice, particularly to emerging economies.

World Bank: Created in the aftermath of WWII to provide loans, particularly to emerging economies, for infrastructure and economic development and poverty reduction.

World Trade Organization (WTO): The successor to the General Agreement for Trades and Tariffs (GATT, also created in the aftermath of WWII) to facilitate trade through its multilateral trading system and provide a venue for resolving trade disputes.

Retail banks: An essential part of the modern economy as these banks hold consumer deposits, facilitate spending (e.g., checking accounts, credit cards), provide consumer loans (e.g., homes, cars), and help expand the money supply through fractional reserve banking, which allows them to lend a portion of the amount held as deposits.

Federal Reserve ("Fed" or central bank): The institution tasked with managing the nation's money supply and regulating the banking sector. In the United States, the Fed increases (to stimulate spending or the economy) or decreases the money supply (to fight inflation or slow down the economy) through the effective federal funds rate (interest rate used by banks to lend money to each other, which is based on the Fed's target range set at one of eight annual Fed meetings), asset purchases (buying bonds or collateralized debt obligations or CDOs, such as mortgages, from banks), and the reserve ratio (the percentage of total deposits each bank must maintain at the Fed).

Other key terms will be in **bold** and defined in the text.

Review Questions

- What is absolute advantage, and how is it different than comparative advantage?
- What are the three main institutions for maintaining the stability of global capitalism and trade? What are the distinctive roles of each institution?
- What is meant by balance of payment? Why is it an important measure of a nation's economic health? What types of economic problems are deficits associated with?
- What are the various roles of the retail banking sector?
- What is monetary policy, and why is it important?
- What is the Federal Reserve, and what are its roles and relationship to the banking sector and monetary policy? How does the Fed manage the money supply?
- What is the role of taxation? What is fiscal policy, and how is it related to taxation?
- What are bonds, who can issue them, and how do they work?
- What are equities, and what role do they play in the financial system?
- What are some other types of investment products in the modern financial system?

Introduction

In the previous chapter, we reviewed some basic micro- and macroeconomic concepts useful for understanding the global capitalist system. We learned that economies are concerned with allocating scarce resources, including labor, to produce goods and services. To this end, most countries in the world have now adopted some variation of market-based capitalism. The market-based approach, which allows people to freely enter and leave the market as either producers or consumers, has ushered in unprecedented levels of production, income, and wealth. The market enables specialization so that labor and resources can be used efficiently to increase the material welfare of society. The tussle between supply and demand helps ensure a dynamism that spurs innovation, leading to an increase in the quality of goods and services offered as well as a reduction in their price over time. In general, this has led to improved standards of living across the planet. Economic inequality, however, has grown rapidly, and both relative poverty in developed countries and extreme poverty in emerging economies continue to persist. Economic growth has also coincided with significant anthropogenic, or man-made, environmental destruction and climate change. As countries grow economically, they emulate the consumption and waste profiles of higher income countries. Thus, more and more natural resources are used to satiate ever-growing demands for energy and products, and more waste is expelled into the environment. Much of this waste is not biodegradable and results in deleterious, or harmful, outcomes for the environment and people. The extent and nature of this impact on public health will be discussed more extensively in Chapter 8.

In this chapter, we delve into the post-WWII international trade system negotiated in 1944 at Bretton Woods, New Hampshire, by the Allied powers, though the USSR was noticeably absent. The conference sought to enshrine free markets and international trade to prevent another global depression. Many of the economic ideas at the conference found their inspiration in the thinking of John Maynard Keynes. We encountered Keynes in the previous chapter as the progenitor of demand-side economics, which argues for active government intervention, when markets are faltering, to stimulate aggregate demand, the total spending by households, businesses, and government. The system created at Bretton Woods includes the (IMF), World Bank, and WTO (which replaced GATT, or the General Agreement on Trade and Tariffs). These institutions play a pivotal role in facilitating international trade, stabilizing faltering economies, combatting poverty, and building infrastructure. **The structural adjustment approach**—opening economies to outside competition and reducing protectionism, government spending, and regulation—used historically by the IMF for economic development, has been criticized by some as draconian and ineffective. The consequence in some cases has been more poverty and inequality. IMF loans were often conditioned on huge reductions in public spending, resulting in a weaker social safety net, a devaluation of the local currency to make exports more attractive, and an opening of local markets to international trade. These austerity measures have eased in the last 2 decades, but some argue that some countries, particularly those in sub-Saharan Africa, require other economic development approaches (2, 9). Chapter 7 explores the legacy of structural adjustment in greater detail.

The demand-side strategies espoused by Keynes were reflected in the Roosevelt administration's New Deal polices and the Obama administration's American Reinvestment Act (ARIA). They were successful in stabilizing the economic freefall after the Great Depression and the 2007–2008 recession, though in the latter downturn, monetary actions by the Federal Reserve, such as quantitative easing (QE), or the buying of bonds and mortgage-backed securities (MBS) from banks, and TARP (Troubled Asset Relief Program, to buy potentially bad mortgages from lenders), also played an integral role in stabilizing the economy. Some have also argued that New Deal policies prolonged the depression or that American recovery wasn't complete until the wartime economy took off during WWII. Others have suggested that the government's actions after the 2007–2008 recession only worsened the systemic problems in finance and banking, set the stage for massive inflation, and will likely result in a bigger crash in the future. To help us better understand the roles of the various pieces of the financial system in the economy, including actions taken during economic downturns, this chapter will discuss the various roles of the Federal Reserve and the banking sector, which provides money to the public, primarily through consumer loans and lines of credit. This chapter will also delve into capital markets, which help companies raise money, primarily through the sale of stocks and bonds.

The banking system is an essential part of the modern economy as it provides a repository for savings and extends credit to consumers using funds on deposit. The trading of stocks and bonds in capital markets plays a similar role for companies, though companies can also borrow from banks. The Federal Reserve, as we learned in the previous chapter, plays a central role in managing the nation's money supply. However, it is also tasked with monitoring and regulating the banking sector to ensure its viability. All three work in tandem to ensure steady economic growth and help steer the economy when it faces strong headwinds and stagnation. The World Bank, IMF, and WTO play a similar role with respect to international trade and economic development, particularly in emerging economies.

Trade: Absolute and Comparative Advantage

As discussed in the previous chapter, specialization, technological growth, economies of scale, and trade are essential features of globalization. To gain a competitive advantage, firms invest heavily in improving product quality and production efficiency by investing in technology and worker skill. Technology- and training-enabled specialization increases the quantity and quality of goods produced, including product consistency, which helps build consumer loyalty. Economies of scale achieved through task specialization and an increase in production volume lower per-unit costs, which are passed onto consumers as lower prices. In short, specialization has helped create the modern world, characterized by an abundance of ever-improving products and services for consumption. Despite increasing resource consumption and waste production, this trend is generally seen as favorable, at least by economists and consumers, as it increases quality of life and the general welfare of the public.

The theory of comparative advantage, described by David Ricardo in 1817 in *On the Principles of Political Economy and Taxation*, argues that specialization should occur for that product or service with the lowest opportunity cost (opportunity cost was coined later, but the concept was implicit in Ricardo's ideas). Ricardo further reasoned that specialization based on comparative advantage and trade would increase the number of goods and services available for consumption (22). Ricardo's theory was an extension of Adam Smith's **absolute advantage** idea, which urged countries to specialize in all goods they could produce efficiently. Economic efficiency occurs when products can be made with fewer inputs (labor, resources, technologies) and only those products that are wanted by consumers are produced (23). The theory of **comparative advantage**, in contrast, holds that total production of all goods and services increases when countries specialize according to their strengths and engage in trade. Countries may have a comparative advantage because of labor, climate, natural resources, technology, or any other factor that reduces input costs and manufacturing time. A high-income country typically has higher skilled labor and better technologies, whereas lower income countries have cheaper labor. Both should specialize according to their comparative advantage, with higher income countries focusing on technology or skill-intensive products and lower income countries on labor-heavy products (this helps explain the outsourcing of manufacturing to lower cost countries like China or Bangladesh and software development in higher skilled countries like the United States). The following example illustrates opportunity cost and comparative advantage.

Comparative Advantage and Opportunity Cost Example

Country	Hours per Pen	Hours per Book	Pens in 100 Hours	Books in 100 Hours
A	1 hr.	4	100	25
B	2 hr.	1	50	100
		TOTAL	150	125
Country	Opportunity Cost*	Opportunity Cost	Pens in 200 Hours	Books in 200 Hours
A	1/4 = .25 Books given up	4/1 = 4 Pens given up	200	—
B	2/1 = 2 Books given up	½ = 0.5 Pens given up	—	200
		Total w/specialization	200	200

Expressed in terms of item given up

Country A can produce a pen in 1 hour and a book in 4 hours. If Country A allocated 200 hours to total production, of which 100 hours are for pens and 100 hours for books, it will produce 100 pens and 25 books in 200 hours. In contrast, Country B can produce a pen in 2 hours and a book in 1 hour. If Country B also allocates 200 hours for total production, of which 100 hours are for pens and 100 hours are for books, it will produce 50 pens and 100 books in 200 hours. Together they will produce 150 pens and 125 books in 400 hours. The **opportunity cost** of pens and books is also shown in Table 5.1. If Country A produces only pens, it only gives up 0.25 books for every pen produced (it takes 4 hours for a book, or 1/4 = 0.25 books per hour). If Country B produces only books, it gives up 0.5 pens for every book produced (it takes 2 hours per pen, or 1/2 = 0.5 pens per hr.). Thus, the opportunity cost of pens is lower in Country A and books in Country B.

If each country focused on the product with the lowest opportunity cost, pens for Country A and books for Country B, and allocated 200 hours to that product only, Country A would produce 200 pens (200 hrs × 1 pen per hour), and Country B would produce 200 books (200 hours × 1 book per hour). By focusing on the items they can make most efficiently, or those with the lowest opportunity cost, and engaging in trade, Country A and B ensure that total world production of pens and books is maximized (200 pens and 200 books versus 150 pens and 125 books). It's important to reiterate that ***the benefits of increased production can only be realized if both countries engage in trade.*** This, in a nutshell, is the basis of globalization. Free trade increases the amount of goods available for consumption and drives prices down as firms specialize and reap the benefits of economies of scale. However, there are casualties, specifically domestic workers and producers who may not be able to compete with cheaper, and sometimes higher quality, foreign goods.

Realizing the importance of trade for both economic prosperity and political stability, as countries engaged in trade would grow dependent on each other, there was a lot of discussion by the Allies on approaches and institutions to prevent another depression and world war. Though many ideas were floated, unimpeded trade figured prominently in all. These ideas eventually germinated the IMF, World Bank, and GATT, which would later become the World Trade Organization (WTO). Though John Maynard Keynes played an important role in the Bretton Woods conference, the main catalyst was the Roosevelt administration, specifically Harry Dexter White, who rejected many of Keynes's ideas in favor of the approaches and institutions that were eventually adopted (24).

The International Monetary Fund, World Bank, and World Trade Organization

The impact of WWII on the political and economic order of the world cannot be understated. Politically, the aftermath of the war gave rise to the Cold War world order. Europe was cleaved with the Union of Soviet Social Republics (USSR) and Warsaw pact countries on one side and the United States and NATO on the other (not every non-Soviet allied country belonged to NATO). The label third world was initially used to describe countries who were non-aligned with the United States and Western Europe, or part of the Soviet fold. Eventually the appellation came to mean poor and underdeveloped irrespective of affiliation. Economically, WWII solidified American economic ascendancy as the United States became the world's manufacturer as Europe was decimated and Japan and China were not yet manufacturing powerhouses. Western European countries would quickly embrace free market capitalism as the United States pumped $13 billion (over $100 billion in today's dollars) into postwar reconstruction through the Marshall Plan (3). U.S. investment wasn't the decisive factor in the post-WWII economic growth and integration. The German economy, for example, grew more rapidly despite France and Britain getting more aid per capita. However, U.S. investment through the plan, coupled with the transfer of manufacturing knowledge and a push for European economic liberalization and political cooperation, was an important factor in Europe's post-WWII economic and political resurgence (3). A reconstituted Europe also provided a market for American goods and a bulwark against the Soviet bloc. As the Ukraine conflict today demonstrates, the disintegration of the USSR was not enough to remove the historic divide between the Russia and the West. NATO continues to exist and will likely admit new member states (e.g., Sweden) given the Ukrainian conflict, bolstering its strength and spurring a new arms race in Europe.

To understand future trajectories, we must often look to the past. In July of 1944, representatives from 44 nations met at Bretton Woods, New Hampshire, to create a new international system for trade. The goals of the meeting were to reduce or eliminate existing barriers to trade, stabilize exchange rates, and stimulate economic growth. The legacy of that meeting is still with us today in the form of the **International Monetary Fund (IMF)**, **World Bank** (then known as the International Bank for Reconstruction and Development), and WTO (previously GATT). International trade and economic cooperation were seen as the touchstones of prosperity and peace. To this end, the IMF was tasked with maintaining the stability of the **global monetary system**, which allows countries to engage in trade. The IMF does this by monitoring the economic policies of its member nations and provides recommendations to resolve risks and instabilities. The IMF develops capacity by providing advice and technical assistance on matters related to fiscal and monetary policy, exchange rates, regulation of the financial system including banking, and legal frameworks (4). The IMF also provides loans and lines of credit to help countries with **balance of payment (BOP)** issues. Currently, the IMF has nearly a trillion dollars to lend to its member countries through its various loan programs and lines of credit (11). Most of its outstanding loans or credit lines are to developing countries, though a few have been extended to middle-income countries such as Chile and Mexico. The IMF is funded by its member countries, which pay dues, or have quotas, based on their economic capacity. The IMF can also borrow additional funds from member countries under its New Arrangements to Borrow (NAB) or Bilateral Borrowing Agreements (BBA) programs (35).

Balance of Payment

The **balance of payment ledger** provides a summary of the economic transactions between a country's residents and nonresidents in a certain period and includes three accounts: current account, capital account, and financial account (7). The current account includes flows of goods and service and primary and secondary income transfers between residents and nonresidents (7); the capital account includes nonproduced, nonfinancial assets and capital transfers, including debt forgiveness, between residents and nonresidents; and the financial account includes net acquisition and disposal of financial assets and liabilities (7). The system uses double-entry accounting with a credit and debit ledger. Each sale of a good or transfer of assets, financial or real, is recorded in the credit ledger and offset by an increase in income payable in the debit ledger for the exported asset or good. For example, if Country A sells a car worth $100 to another country, Country A would record $100 (CAR) in the credit ledger and $100 as income payable in the debit ledger. The two ledgers cancel each other out (i.e., $100 – $100 = 0), so the balance of payment is theoretically, and ideally, zero.

In reality, the ledgers are rarely zero, and many countries have balance-of-payment surpluses or deficits. According to the U.S. Bureau of Economic Analysis (BEA), the BOP current account deficit for the United States was $291 billion for the first quarter of 2022, up from $225 billion in the fourth quarter of 2021 (8). The BOP deficit in Q1 2022 was 5% of the U.S. GDP. Large BOP deficits can occur for different reasons; however, they usually don't portend well for the economy when imports are substantially more than exports (i.e., countries buy more than they sell) or if a government is unable to pay its debts (treasuries or bonds). Large deficits can devalue currency and cause capital (money) flight as wary investors and their money leave the country for safer pastures. Even a large price increase of imported commodities (e.g., food or oil) can trigger a balance of payment crisis as countries must continue to import these essentials without passing on the increase in costs to consumers, who are often too poor pay for the price increases. Governments must either print more money, further devaluing their currency, or use their foreign currency reserves (i.e., money from another nation, such as the dollar that is widely accepted as a method of payment) to pay for the needed imports. Ongoing deficits in economically weaker countries can lead to an economic crisis resulting in insolvency (bankruptcy) or intervention by external actors, usually the IMF, which issues emergency loans with strict requirements. The IMF has been criticized by many for adopting policies not appropriate for many developing countries. These include **capital market liberalization** (i.e., allowing money to come in and leave the country easily), tight monetary policy (i.e., reduced printing of money), and **trade liberalization** (i.e., allowing foreign products in local markets). These policies, sometimes called the **Washington consensus** will be discussed in Chapter 7. Joseph Stiglitz, former World Bank chief economist and a Nobel Prize winner in economics, has likened the Washington consensus policies to setting sail in a boat that hasn't been repaired, without life preservers, and whose crew lacks the training to navigate the rough seas (9). Stiglitz argues that IMF and World Bank policies have, in many instances, worsened the economic and political circumstances in poor countries that came to them for help (9).

The United States' unique position as the world's primary reserve currency allows it to consistently maintain a large balance of payment deficits. A **reserve currency** is one that is held by other countries as part of their foreign exchange reserves because it is commonly accepted throughout the world as payment for goods and services. Central banks and other financial institutions throughout the world maintain large U.S. dollar reserves as the dollar is accepted as payment for anything, legal or otherwise, on the planet. This creates a constant demand for the U.S. dollar. Secondary reserve currencies include the Euro (EU), yen (Japan), and yuan (China), which are also used in financial transactions worldwide and held as reserves. None, however, including the Chinese yuan, have thus far managed to successfully supplant the dollar; this may change in the coming decades given the economic and political realignments underway due to the growth in the Chinese economy and world events such as the Ukraine conflict, whose geopolitical impacts are far from certain. The U.S. dollar reserve status stems from the Bretton Woods conference, when it was chosen as the world's **numeraire**, or benchmark, currency. The dollar's reserve status was officially enshrined in the IMF's Articles of Agreement in 1945 (5). To stabilize exchange rates, countries agreed to keep their currencies within 1% of the dollar, and the dollar was indexed to gold at $35/ounce. The United States became responsible for both keeping gold prices fixed and maintaining a dollar supply that allowed convertibility into gold (5). In other words, the linking of U.S. dollars to gold limited the United States' ability to print dollars.

This situation became untenable by the latter half of the 20th century as U.S. balance of payment deficits forced the United States to abandon the gold standard under the Nixon administration in 1971. This opened the door for **fiat** currency, or currency that is not backed by a commodity such as gold. The value of fiat currency derives from the willingness of others to accept it as payment (i.e., supply and demand), which, in turn, depends heavily on the perceived economic and political stability of the issuing country. Countries with weak economies and political instability that print currency excessively usually experience significant inflation. Zimbabwe is often touted as the poster child of hyperinflation. Land reforms in 2000–2001, which reallocated farms from White commercial farmers to Black families, coupled with persistent droughts, slashed agricultural productivity by over 50% between 2000–2009 (6). The government also printed copious amounts of money to pay for war-related efforts and to settle external debts. Thus, by 2008–2009, the country, beset by years of a declining economy and soaring public debt, experienced a 17% contraction in GDP, 94% unemployment, and 231,000,000% annual inflation (6). The IMF had stopped making loans to Zimbabwe before the new millennium and adopted a policy of noncooperation with the Mugabe government. In light of the deepening economic crisis, however, it agreed to a $510 million loan in 2009. More recently, in December 2021, the IMF denied further debt relief to Zimbabwe unless it paid off its World Bank loan (Zimbabwe had already settled its IMF loan).

The **World Bank**, whose genesis can also be traced to Bretton Woods, focuses on poverty alleviation and economic development. The World Bank is comprised of five organizations, collectively called the World Bank Group: the International Bank for Reconstruction and Development (IBRD), International Development Association (IDA), International Finance Corporation (IFC), Multilateral Investment Guarantee Agency (MIGA), and International Center for Settlement of Investment Disputes (ICSD) (14). The IBRD makes loans to middle-income and economically stable lower income countries, whereas the IDA focuses on emerging economies without a mature financial sector. The IFC seeks to stimulate private sector investment in developing countries and provide capital in the form of loans or equity. MIGA, like the IDA, seeks to increase private sector investment by providing insurance and credit enhancements, which reduce the perceived risks of capital investment in emerging economies. The ICSID acts as a mediator for investment-related disputes (14).

The World Bank funds development projects in education, health, public administration, infrastructure, financial and private sector development, agriculture, and natural resource management. In many cases World Bank funding is combined with other public and private sources (12). Like the IMF, the World Bank is funded by contributions from its members. In 2022, the World Bank had over $320 billion invested worldwide in 3,017 active projects in 137 countries. The African subcontinent is the biggest recipient of World Bank funding ($110 billion), followed by South Asia ($63 billion). Between $30–$40 billion has been allocated to each of the following sectors: energy, transportation, social protection, water and sanitation, public administration, and health (13). A particular sector sometimes focuses on issues that are also targeted by other sectors. COVID-19-related programs, for example, have been an important area of activity the past 2 years: The World Bank Group contributed $157 billion to COVID-related efforts during the last 2 years; and COVID-related programs can be found in the health, social protection, and public administration sectors (14). World Bank–funded programs help provide basic necessities such as food and medicines to the most vulnerable and high-tech innovations, such as the expansion of digital education programs. Aid also helps strengthen financial systems to expand access to credit, capital, and jobs. Many in low-income sectors in developing countries cannot invest in themselves or their businesses because they lack access to debt (i.e., the ability to get a loan), equity (i.e., capital investment from others), and insurance to provide a buffer against the unexpected. In contrast, retail and investment banking services, access to credit (i.e., the ability to assume debt), and insurance are taken for granted in developed countries. Though access often varies depending on an individual or household's **socioeconomic status (SES)**, or income and wealth, most developed economies have some access to these financial services. In developing countries, these services are usually limited to the middle and upper SES (Socio-Economic Status) groups, with large swaths of those in the lower SES groups, particularly those working in the informal and unregulated sectors, without any access to the tools that promote upward economic mobility. Bringing these services to the poor and lower SES groups has been a central challenge of national governments and the IMF and World Bank. The banking and financial sectors are discussed in greater detail later in this chapter. Additionally, the World Bank's economic development and poverty reduction programs will be discussed in Chapter 7.

Globalization has helped make trade an essential part of the world's economy. For example, during the first quarter of 2022, $5.8 trillion in merchandise was exported throughout the world (17). In April 2022, the United States exported $252 billion in goods and services and imported $340 billion for a deficit of $87 billion (18). Ensuring the smooth flow of goods and services across national borders was an important goal of the Bretton Woods attendees. Their discussions and negotiations during the conference eventually led to the adoption of the General Agreement for Trades and Tariffs (GATT) in 1947. GATT was created to facilitate international trade by reducing or eliminating barriers such as quotas and tariffs (taxes). In 1995, it was replaced by the **World Trade Organization (WTO).** Whereas the IMF and World Bank were created to address macroeconomic policy, economic development, and poverty alleviation through the provision of technical assistance and loans, the WTO was designed to promote international trade through its **multilateral** (i.e., involving more than three countries) trading system, at the center of which are its world trade agreements. The WTO helps its members develop, implement, and interpret these agreements and, if necessary, arbitrate trade disputes within the context of these agreements. Though often lengthy and complex, the trade agreements promote transparency, fair competition, reduced trade barriers, collaboration, and environmental protection (15). Goods, services, and intellectual property are covered in the agreements.

The **WTO** has a process for settling trade-related disputes amongst its member nations. Members can bring disputes they are unable to resolve themselves for review by the WTO. This process is incremental and can lead to arbitration by an expert panel. In addition to settling disputes, the WTO monitors trade and trade policy among member countries and provides technical assistance to increase the trade capacity of emerging economies (15). Today, the WTO's 164 member countries account for nearly all (98%) of the world's trade. The WTO's explanation of the importance of world trade is succinctly captured in its most recent (2021) annual report, which notes "today's hyperconnected global economy, characterized by deep trade lines, has made the world more vulnerable to shocks, but also more resilient when they strike" (15). The interconnectedness of the world was evident during the early days of the 2020 pandemic. The movement of people and animals; urbanization and population density increase; and deforestation and environmental degradation, including climate change are all implicated in the spread of zoonotic illnesses such as COVID-19 (16). Additionally, trade is implicated in technological shocks (e.g., cyberattacks), industrial and economic shocks (e.g., supply chain crises), and sociopolitical shocks (e.g., conflicts, political stability). This acknowledgement of the potentially severe negative impacts of trade is also an indictment of globalization and the risks it creates, particularly in more vulnerable, emerging economies. The WTO's report, however, also argues that these shocks are not inevitable, and as much as trade increases the risk for certain shocks, it also provides the tools to address the fallout if countries act collaboratively, transparently, and not only with their interests in mind. Unfortunately, the basic tenets of capitalism emphasize individual utility, which countries espouse in their policies (e.g., America First) that prioritize national interest over all else.

In addition to the World Bank, IMF, and WTO, there are several other multilateral organizations formed around international trade. The **Organization for Economic Cooperation and Development (OECD),** for example, was formed in 1961 and today includes 38, mostly higher income, countries. The OECD members collectively account for nearly 80% of the world's trade. The **OECD** is primarily an advisory body that convenes experts and policymakers to foster dialogue, set standards (over 450 international standards have been developed through its activities), and formulate policy. The areas covered by the OECD are broad and include economic growth and development, taxation, regulation, public health, human development, education, and the environment. The OECD provides analysis and advice to member countries, as well as the **G7 and G20**, an informal group of economically powerful nations. The G7 nations include Canada, the United States, Italy, German, France, the UK, and Japan; the G20 includes all G7 nations plus Russia, Argentina, Australia, Brazil, China, India, Mexico, Saudi Arabia, South Africa, South Korea, Indonesia, Turkey, and the European Union. The G7 is older and has met regularly since the mid-1980s, whereas the G20 started convening in 1999. Both organizations meet to discuss important economic and political issues, though the G20 has traditionally focused more on economic issues as, unlike the G7, not all members are full democracies.

Some have argued that the G7 is growing irrelevant as the economic center of gravity, as measured by GDP growth and energy use, is shifting to the G20, particularly the BRIC (Brazil, Russian, India, and China) countries (20). Others have argued that the political incompatibility of the G20, particularly the BRIC countries, and the increasing isolation of Russia, which has increased due to the Ukrainian war, make cooperation difficult. Additionally, China has emerged as a bona fide political and economic competitor to the United States and become more **statist** or centrally controlled by the Communist party; additional strains are created by China's aggressive posture in the South China Sea; its Belt and Roads initiative, which some in the West see as a thinly veiled attempt to create vassal states; and refusal to recognize Taiwan's independence (21).

The World Economic Forum (WEF) is another organization without any formal role that nonetheless attracts the rich and powerful to its annual meeting in Davos, Switzerland. The first meeting was held in 1974; today the annual WEF meeting is an invitation-only event that draws billionaires, heads of state, and economic, political, and other luminaries. Some critics of the forum allege that attendees simply gather and strike business deals; others insinuate more sinister motives of world domination and control. Some have argued that the WEF, though elitist, serves a valuable function as it provides a venue for the political and economic elite to discuss a range of issues, from the economy to climate change, and informally agree on a course of action.

Financial System: Banks and Stocks

Opening a bank account is a rite of passage and marks an important steppingstone to adulthood in developed and middle-income countries and for the middle and upper middle class in emerging economies. For the general public, retail **banks** are safehouses for capital (money) or lenders for important consumer purchases such as homes or automobiles. Banks also issue credit cards to provide access to short-term loans for everyday consumer purchases. Retail banking is essentially the nexus between depositors and borrowers. Those depositing money into banks provide the capital used for bank loans. There was a **run on banks** during the Great Depression (1929–1939) as most depositors sought to withdraw their deposits as they were unsure if the banks would remain **solvent**, or in business. This "run" bankrupted the banks. To prevent such a run from occurring again, the U.S. government now insures deposits up to $250,000 per ownership category through the Federal Deposit Insurance Commission (FDIC) and scrutinizes bank health through multiple regulators, including the Federal Reserve. Typically, only a small proportion of depositors withdraws funds at any given time. This allows banks to use deposits to make loans and extend credit. This capacity to use a portion of deposited money for loans and credit is called **fractional reserve banking**. Banks, thus, essentially create money by lending a portion of their deposits, though depositors such as you and me still see the full amount of our deposits on our monthly account statements. Banks **underwrite** loans and credit, or assume the risk associated with loans and credit, by assessing borrower capacity to **service** the loan by making regular payments towards interest and principal. In the United States, a credit score, which varies between 0–850, provides a snapshot of borrower credit worthiness. Banks use the credit score, along with other consumer data (e.g., income verification), to make lending decisions.

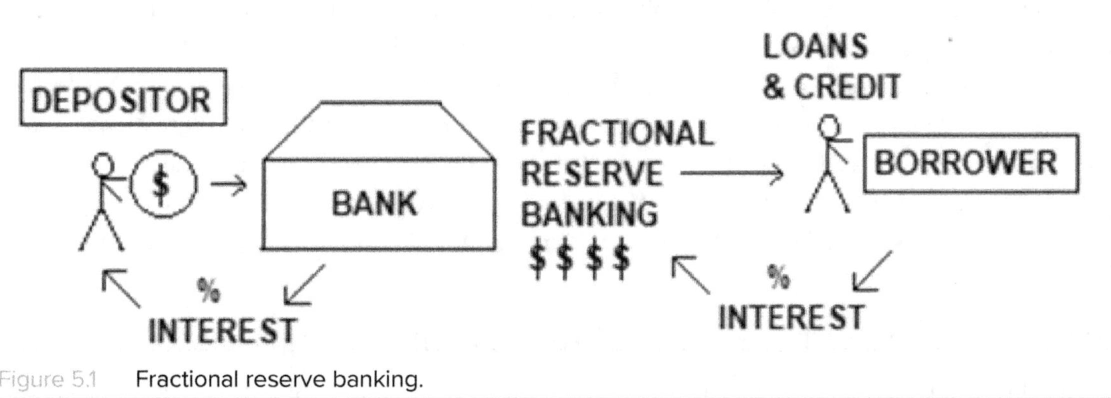

Figure 5.1   Fractional reserve banking.

This process is integral to the economy as banks provide capital to spur consumer spending on small and large consumer items such as automobiles and homes. The availability of loans and credit allows consumers to purchase essentials (e.g., food) and engage in discretionary spending (e.g., goods that are desired but not needed). Loans and credit also allow consumers to invest in themselves, as loans can be used for small businesses, investments, education, health care, and other goods and services. The lack of access to **credit** and **insurance**, a contract that provides payment for an adverse event in exchange for a premium, is one of the biggest factors impeding upward economic mobility in emerging economies and will be discussed in Chapter 7. Insurance companies use the combined premiums of those paying insurance to pay those who experience adverse events; if the total amount paid is less than the total premiums received, the insurance scheme is profitable. In some cases, widespread catastrophic events can cause insurance bankruptcies. In the U.S. real estate market, widespread use of mortgages and insurance has helped expand homeownership, protect homeowners, and create wealth as buyers can leverage a smaller equity contribution (i.e., down payment) of usually 20% to purchase a more expensive home with the remainder financed by a home loan, usually a 30-year amortizing mortgage in which regular monthly payments include principal and interest. Also, the monthly payment does not change during the **life of the loan** (30 years, or 360 months), though interest is front-loaded (i.e., the interest payment is maximum during the 1st month and gets progressively smaller with each subsequent payment, though the total monthly payment stays the same over the life of the loan).

In the United States, an 80/20 LTV (loan-to-value) mortgage structure, in which the buyer provides 20%, typically as down payment (equity) and the bank provides 80% as a mortgage (debt), is common. This structure was instrumental in expanding homeownership in the aftermath of the Great Depression. Though amortizing mortgages were used in the United States before the 1930s, the LTV ratios were low, and maturity periods were 15 years or less (27). The U.S. government further incentivized homeownership through the FHA insurance program in 1934 and Serviceman's Adjustment program in 1944, which provided federal guarantees for home loans. **Guarantees** are important as loan default risk (i.e., risk that certain buyers would not pay their mortgages) is assumed by the government, which allows for expansion of the consumer home loan market. Fannie Mae (1938) and Freddie Mac (1970), two government-sponsored enterprises (GSE), helped spur the secondary mortgage market by buying mostly single-family mortgages from banks and other lenders. Today these two GSEs either own or guarantee nearly half of the entire U.S. mortgage market. In their absence, the U.S. home mortgage market would be less robust, with fewer homeowners and slower price appreciation. Though Fannie and Freddie are private entities, they are "**too big to fail**" as their failure would likely collapse the U.S. real-estate market. Given their importance, they are, *de facto,* backed by the U.S. government. Banks that make the initial loans and **underwrite**, or evaluate, prospective buyers are the **primary lenders**. Many sell their loans on the **secondary markets** to Fannie or Freddie and other private investors that **recapitalize** them, allowing them to continue making loans to new prospective home buyers. The secondary market for home loans thus allows homeownership to thrive. Other important U.S. government polices to support homeownership include the tax deductibility of mortgage interest and property taxes (tax deductions were discussed in Chapter 4), depreciation (i.e. a tax deduction) of investment properties, and capital gains exemptions for primary homes (24). The U.S. housing market and its role in building wealth will be discussed more extensively in Chapter 7.

Bonds, Stocks, and Other Financial Instruments (Securities)

Homeownership is not the only arena to use debt to drive market expansion. Debt plays a central role in capitalism as it allows individuals, companies, and the government to use external sources of capital to fuel growth or support spending. Consumers use debt in the form of loans and credit for consumer purchases (e.g., clothes, homes, automobiles, self-investment [e.g., education, health care], and financial investment [e.g., small business loans]). Companies use debt for operations and expansion. Companies secure debt through loans, lines of credit, and **bonds**, which can be sold publicly (i.e., on a stock exchange) or privately (i.e., directly to investors) (28). Bonds have a **face value**, the amount a bond can be redeemed for after it has matured (i.e., loan period), and a coupon rate, the amount of interest paid on the face value at some regular interval. For example, a 2% semi-annual coupon rate on a bond with $100 face value with a 10-year maturity rate would result in $2 paid to the borrowed every 6 months for 10 years along with $100 at the end of the 10-year period. Since bonds can be traded on secondary markets, the resale value of bonds may sometimes exceed the face value depending on the present value of alternative investments (28).

Governments can also issue bonds. In the United States, cities (municipal or "muni" bonds), states, and the federal government issue bonds, as do school districts and other entities (e.g., infrastructure project companies or nonprofit organizations) (28). The U.S. bond market is the world's largest and was estimated at around $46 trillion in 2020 (29). Like any other investment, bonds carry risk: In the United States, corporate and municipal bonds are typically riskier than state or federal bonds, which are also called **Treasury Bills (T-Bills).** The U.S. government has issued trillions of dollars in bonds (debt) to finance spending. As of April 2022, nearly $7.5 trillion in U.S. **treasury securities**, or bonds, was held by foreign governments. The biggest holders today are Japan ($1.2 trillion), China / Hong Kong ($1.2 trillion, or $1.43 trillion if Taiwan is included), the UK ($612 billion), Ireland ($309 billion), and Luxemburg ($294 billion) (30). The U.S. dollar's reserve status makes U.S. bonds very attractive to foreign governments and other institutions as a U.S. **bond** is essentially a guarantee by the federal government that it will pay the bond's interest rate in dollars and exchange the bond for actual dollars once it matures. The attractiveness of U.S. bonds is a major factor in allowing the United States to consistently run large deficits without a significant devaluation of its currency. No other country can run the same deficits. This is one of the reasons China, the EU, and others are seeking to create alternatives to the U.S. dollar for international trade and finance.

Companies can also raise money by allowing investors to purchase equity (i.e., shares or stocks) in the company, privately or publicly, through the purchase of stocks on an exchange like the New York Stock Exchange (NYSE). Most large companies in the world are publicly traded, with stocks listed on a major stock exchange. This allows companies to raise large amounts of capital and investors the opportunity to profit from the appreciation of stock over time. The swing in stock prices depends on how attractive a stock is relative to other investments. Companies that demonstrate growth in their quarterly reports and long-term growth potential usually have stocks that appreciate. Stocks from different companies can be bundled together in different packages and traded. These bundles include mutual funds, exchange-traded funds (ETF), or another type of **equity security**. These are marketed and sold as investments by **brokers, investment bankers, and financial advisors** to individuals or institutional investors such as managers for pension or retirement funds.

The names used to describe the bundled stocks depend on their underlying characteristics (e.g., type of stocks in the bundle, trading frequency, etc.). Similarly, consumer loans for homes, automobiles, and other products can be bundled together and sold to investors as a financial product (e.g., **collateralized debt obligations**). A security refers to any type of investment in a company and can include stocks (equity security), bonds (debt security), or limited partnership interest (equity security) (31). In the United States, the buying and selling of securities is regulated by the Securities Exchange Commission (SEC). The financial services industry has developed a number of complex securities, some of which are extremely risky and not appropriate for retail (i.e., average) investors. These include various types of **derivatives**, or contracts, whose value depends on some underlying asset such as stocks, bonds, mortgages, and consumer loans (CDOs); currencies or commodities; fungible goods such as agriculture crops (i.e., can be exchanged for other goods of the same type); precious metals; or energy (oil and natural gas) (32). Derivates allow investors to **hedge** or spread investment risk through diversification of an investment portfolio and profit from changes in currency exchange rates. Derivates also allow investors to leverage smaller cash investments as contracts cost a fraction of the underlying asset on which they are based. Derivates include swaps, futures contracts, options, and a number of other sophisticated financial instruments (32).

Innovations in the financial industry have allowed investors to profit from rising equity, commodity, or security prices and price decreases (e.g., **put options**). The innovations in the financial services industry have increased both the upside or profit potential and downside or risk. The systemic risk (i.e., risk to the entire financial system) from complex financial instruments cannot be understated. The 2007–2008 recession was in large part due to the excessive use of complex financial instruments. Lehman Brothers, a major investment bank, collapsed almost overnight given their large exposure to such instruments. The contagion, like a biological pandemic, spread from the U.S. housing markets to the broader U.S. economy, eventually causing a recession that spread to the world as trade decreased and countries became unable to pay their bills (i.e., **sovereign debt crisis**).

Since the great recession, regulators throughout the world have taken important steps to reduce **systemic risk**, or threats to the larger economy, by adding new regulations to reduce the risks taken by banks and financial institutions and creating firewalls to contain the spread of a collapse in a particular financial sector. Regulation is necessary because capitalists often take risks that undermine the very system that allows them to profit. These risks have created a financial services sector that is estimated to exceed $20 trillion today and involve companies, individuals, and virtually all other types of institutions that have exposure to financial markets through retirement accounts, investment accounts, insurance, and loans. This growth has made those with exposure to the market wealthy, particularly in the last 10–13 years as markets have been on a **bull** run, or growing consistently (a **bear** market, in contrast, is one that is shrinking). Those with less exposure, usually the lower SES groups in developed countries or majorities in poor and emerging economies, have been left out of the windfall. This is a major cause of the growing economic inequality (i.e., growth rate of capital > growth rate of incomes) according to some economists (e.g. Picketty) and will be discussed in Chapter 7 (34). Ultimately, governments must help foster economies that not only grow but also share the dividends of that growth equitably.

Taxation

One of the ways the government redistributes income and wealth is through **taxes**, the proportion of income (paid for **labor**) or wealth (**capital gains or profit on investments**) collected by the government. The government can also use the tax system to implement its stimulatory **fiscal policies** as tax credits and deductions can incentivize certain behaviors such as the purchasing of solar panels, electric vehicles, or homes. In the United States, taxes are levied by different levels of government (municipal, county, state, and federal) on various types of activities and incomes, including property, income, capital gains, and sales. The income and capital gains taxes are the biggest sources of federal and state government revenue in the United States. The development and implementation of an effective tax system is a balancing act between the distortion-inducing effects of taxes and the public goods for which taxes are used. In other words, taxes reduce market efficiency but also provide funding for spending in a myriad of areas, such as defense, infrastructure, social services, housing, environmental protection, and corporate bailouts. Income taxes are often **progressive**: One pays a higher proportion of income (i.e., higher tax rate) toward taxes as income rises; other taxes such as those on fuel can be **regressive** and impact lower SES groups disproportionately as they often work in jobs that are not telecommuting friendly and the taxes represent a larger share of their income. A major debate in the United States and other countries is the taxation of wealth, or taxes on **unrealized capital gains** on assets that have appreciated in value (e.g., equities or real estate) but have not been sold. The wealthy can use these assets to secure debt for spending without incurring a taxable event (i.e., they can get a line of credit against their house to buy things without paying taxes on the home equity that is the collateral or basis for their line of credit). As home prices and stocks grow, their wealth grows along with their ability to use their wealth to procure loans, often with interest rates far below the rate of return on their investments, for regular expenditures.

The Federal Reserve

As discussed in the previous section, a healthy financial services sector is essential for economic growth and stability. The government must strike a balance between innovation in the financial services industry and the attendant risks posed by innovation, especially those to the broader economy. The Federal Reserve, or central banking system, which has been adopted worldwide, plays a central role in regulating the banking sector and managing the money supply, or **monetary policy**. The U.S. Federal Reserve system is comprised of 12 geographically specific reserve banks with 24 branches. The reserve banks perform several regulatory functions: Supervise and assess state member banks; lend to depository institutions (to ensure bank liquidity for consumer lending); distribute coin and currency to banks, clear checks, and manage the ACH system; and examine financial institutions for compliance with consumer protection and fair lending laws (25,26). The reserve banks also provide regionally specific economic data to the system's main decision-making body for monetary policy: the Federal Open Market Committee (FOMC), whose members include the seven board of governors, each appointed by a U.S. president, and five reserve bank presidents. The Federal Reserve impacts the money supply through three principal means: the federal funds rates, reserve ratio, and open market operations (25, 26).

The **reserve ratio**, or the percentage of total bank deposits that must be held at the Federal Reserve, is another important mechanism for increasing or decreasing the money available for loans and credit at banks (a higher reserve ratio requirement means more money must be held at the Federal Reserve, leaving less for lending). Since March 15, 2020, the reserve requirement has been zero to stimulate lending during the economic slowdown during the early days of the COVID pandemic (26). Prior to the zero rate, the Federal Reserve used a tiered system in which deposits below a certain amount were exempt from the reserve requirement, deposits between the exempt amount and the "low reserve tranche amount" were subjected to a 3% requirement, and 10% reserve was required for deposits above the "low reserve tranche amount" (26).

The **federal funds rate** is the rate paid on reserve balances at the Federal Reserve. **Depository institutions**, such as banks, savings and credit unions, and government-sponsored enterprises (GSEs), such as Fannie Mae, borrow from each other for short-term capital needs on a regular basis. The interest rate they charge for short-term loans is unlikely to be less than the federal funds rate because a depository institution could borrow money from another bank and deposit it at the Fed for a higher interest rate, capturing the difference in the lending rate and federal funds rate as profit. Thus, the federal funds rate becomes a **benchmark** index for other consumer lending rates: The higher the federal funds rate, the higher the consumer rates. The lowering and increasing of the funds rate allow the Federal Reserve to increase or decrease the cost of consumer credit and loans, thereby increasing or decreasing the capital or money available in the economy. In 2022–2023, the Fed raised rates several times to combat entrenched inflation.

The Fed raised its funds rate by 400 basis points, or 4% in 2022, and the increases continued in 2023, though inflation reduced considerably during the year. The federal funds rate increases are designed to reduce inflation. As interest rates rise, lending declines, and, by extension, spending (i.e., less dollars chasing goods and services means reduced demand, which should, theoretically, reduce prices).

The Federal Reserve can also increase the capital available for lending though its purchase of U.S. treasuries and mortgage-backed securities (MBS) from banks and GSEs (e.g., Fannie Mae). By buying bonds and mortgages, the Fed increases liquidity, or the amount of capital in the financial system, as the bonds and MBS held by banks and GSEs are exchanged for money. The increase in capital should, theoretically, increase lending by banks and GSEs, creating a stimulatory impact on the economy. This monetary policy mechanism was used extensively in the aftermath of the 2007–2008 recession to reinvigorate lending and, by extension, economic growth.

Another important function of the Federal Reserve is to evaluate the viability of depository institutions in various economic conditions. The Fed does this by using criteria defined in the Dodd–Frank stress test and capital buffer test, which evaluate the institution's capacity for maintaining adequate liquidity (availability of cash) under different scenarios, especially during periods of economic downturns when the public is most likely to increase withdrawal of its deposits (25).

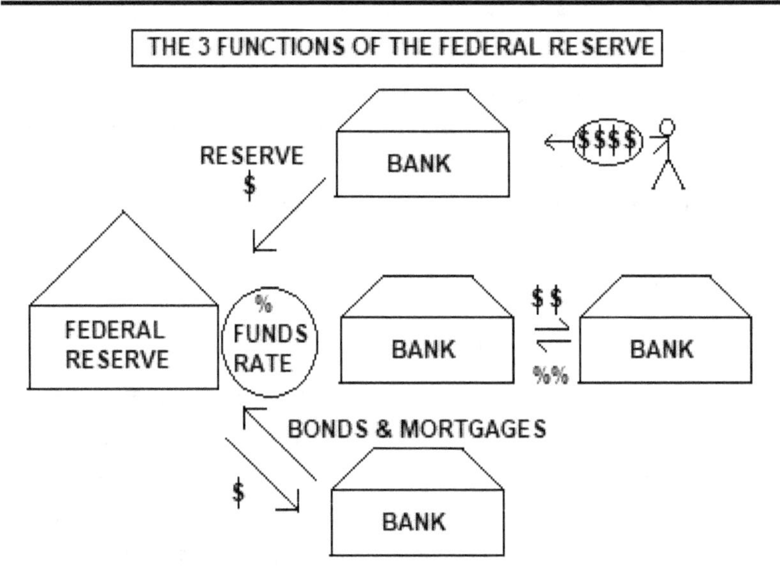

Figure 5.2   Three functions of the Federal Reserve.

Conclusion

The preceding sections have outlined the basic financial systems developed during the 20th century to stimulate and regulate national economies and world trade. The systems have grown increasingly complex; however, there remains an unyielding faith in the ability of people to improve the general welfare of all society through specialization, interdependency, and trade—the tenets of global capitalism. Furthermore, recessions, depressions, and wars are seen as anomalies whose impacts can be mitigated and contained through the application of the appropriate analyses and interventions. This is the central task of government and the institutions of international trade and development: IMF, the World Bank, and the WTO. Climate change and sustainability-related issues offer another set of challenges. Nonetheless, there remains a belief in the ability of humans to best all that is thrown at them. This is a central belief of enlightenment thinking: Increasing reason and application of the **positivist** methods (i.e., scientific method) can and will trump all adversity, including the differences that separate people, cultures, and nations from each other. However, is this optimism warranted? Is the Enlightenment-based global capitalist model sustainable? The Enlightenment is a project of Western civilization. Its promise, or some would say hubris, is the creation of a world culture based on reason, positivism and, later, adoption of universal suffrage and extension of civil rights to many types of previously marginalized groups. The Enlightenment, as the word evokes, was a radical shift in **epistemology**, the process by which we *know* things. The scientific method, or positivism, replaced the older ways of knowing things and imparting causality in which divine and supernatural forces played a decisive role in human affairs. Instead, hypotheses about the world were tested using knowledge derived **empirically**, through observation and collection of data. The Enlightenment project has continued apace, imbuing its ideals into every sphere of human activity from the economy to politics and technology.

The Enlightenment created the modern individual, free and unfettered from the chains of tradition, community, history, and place. In other words, the universal human rights championed by the Enlightenment created the type of individual necessary for the growth of market-based economies and globalization. Only individual and groups, free to exercise their agency, or will, in a market to maximize their utility, or happiness and satisfaction, can create the global-capitalist system being exported wholesale to countries across the planet (the United States, as some argue, is not a hegemon like the empires that preceded it; it is the biggest proponent or cheerleader of this model; its many interventions across the planet in defense of democracy and nation building are geared specifically at creating a human cultural based on Enlightenment values). However, the question remains, are these universal human values, and can they effectively supplant local cultures? And, if so, what is the cost? The evolution of different types of capitalism in Russia, China, and other parts of the world; the rejection of certain types of human rights (e.g., women's, LGBTQIA); and the persistence of religion- or ideological-based conflict seem to indicate that Enlightenment's ideals are far from universal.

The more dour analyses, such as those offered by philosopher John Gray, would emphatically reject the idea of a global human culture. Gray argues that markets and global capitalism inveigh against tradition, localism, and community. Growth of the market has coincided with the destruction of local traditions, created vast inequalities, and destroyed the environment through pollution, resource consumption, and climate change. Is global capitalism indeed the panacea its proponents claim? Can it be rehabilitated, and, if so, what does a reconstituted global capitalism look like? Are the answers within global capitalism itself? If not, what are other ways of organizing production, consumption, and human society? There are no simple or easy answers, but a step in the right direction would be to seriously consider alternative ways of organizing human society. That is, we don't have to reject all of global capitalism, but if we are to address its serious shortcomings—and there are many—we must be willing to leave its "matrix."

Chapter 6

Demography

The Study of Population Characteristics and Their Relationship to Important Social Phenomena

Key Terms

Total fertility rate (TFR): The average number of children a woman can be expected to have as she goes through her childbearing years, typically 15–44. The TFR provides a measure of the number of children women are having on average. Replacement-level fertility, the number needed for a population to reproduce itself in the absence of immigration, is usually 2.1 children per woman.

Mortality rate: The number of deaths in a region in a specific time period divided by the population of the region. This is also known as the crude death rate. Mortality rates can be estimated for diseases (disease specific mortality rate) or specific ages (age specific mortality rate).

Life expectancy: The additional number of years a person can be expected to live at every age if the age-specific death rates for a given year prevailed for the rest of the person's life. Life expectancy at birth is the additional number of years a newborn can be expected to live. Women usually have longer life expectancies at all ages than men.

Age structure: The number of people in each age (1, 2, 3, etc.) or age range (0–4, 5–9, etc.) by sex (M, F). Age structure is sometimes called a "population pyramid."

Dependency ratio: The number of dependents, or those below 15 and over 65, divided by those of working age (between 15 and 65). The dependency ratio provides insight into the burden imposed on those who are working by children or the elderly. In some countries, like Japan, falling fertility and high life expectancy has increased old-age dependency, which means there are increasingly fewer people paying into pensions and health care schemes to care for the elderly.

Migration: The movement of people from one area to another. People leaving an area emigrate, while those coming into an area immigrate. Immigration into a nation is easier to predict as it can be controlled through immigration policy. Emigration is more difficult to estimate as it depends on social, economic, cultural, political, demographic, and other factors.

Other key terms will be in **bold** and defined in the text.

Review Questions

- What is total fertility rate (TFR), how is it determined, and why is it important? How is fertility different from fecundity? Why is the TFR falling in many places?
- What is life expectancy, how is it determined, and why is it important?
- What is age structure, how is it represented, and what is its relationship to the dependency ratio? What does the dependency ratio tell us about the future of pension, insurance, and other similar programs?
- What are the different ways of measuring death or mortality?
- What are headship rates? What explains declining headship rates among young people in the United States?
- What are the different periods identified by demographic transition theory? In which periods is the population likely to increase or decrease, and why? Which countries are currently in different periods?

Introduction

The global population crossed 8 billion people in the latter half of 2022 (3). Seven countries account for half of the planet's population. The biggest of these, China (1.43 billion) and India (1.42 billion), exceed the population of the other five: the United States (338 million), Indonesia (276 million), Pakistan (236 million), Nigeria (219 million), and Brazil (215 million) (2). The world's population is expected to stabilize sometime around the latter half of this century with between 10 and 11 billion people (3). Population growth in a region depends on the difference between the absolute number of births and deaths. These are, however, sometimes difficult to predict as many different factors impact **fertility**, the number of children an average woman can be expected to have, and **life expectancy at birth**, the number of years an individual can be expected to live at birth. The demographic experiences of countries are quite different depending on the level of economic development and social factors such as women's rights. The higher population growth in many lower income countries is the result of larger numbers of younger people relative to the elderly (i.e., a younger age structure), higher infant and child mortality, and less control exercised by women over their reproductive health. In contrast, many higher income countries are experiencing population declines due to falling fertility and higher life expectancy. The Japanese population, for example, is expected to shrink by tens of millions in the coming decades. In 2021, Japan had over 800,000 births and 1.4 million deaths for a net population loss exceeding 600,000 (4). This was the largest **natural decline**, measured by the net population decrease when deaths exceed birth, in Japan's history. Japan's fertility has been declining for many years and is now around 1.3, one of the lowest in the world (4). Higher labor force participation by Japanese women, coupled with the persistence of traditional norms that place a higher childrearing burden on women, likely plays an important role in the decline of marriages and children. This creates a "double duty" as women are expected to tend to the household despite working; many, not surprisingly, choose to forgo marriage and/or children. In 2020, about one in seven Japanese men and women, between ages 17–34, stated not wanting to ever get married; 38% of households are single person; and only one in five married households had children (30). Similar trends are occurring in other Asian countries such as South Korea and China (30).

Though the world's population is still growing in absolute terms since the first billion in the 1800s, the population growth rate has slowed considerably and dropped below 1% in 2021 due to declining fertility. In 2022, the average global fertility was around 2.3 children and is expected to fall to replacement level (2.1) by midcentury (3). Lower fertility, coupled with increasing life expectancy due to better nutrition and medical care, has resulted in an older **age structure** (the number of people by sex in different age ranges). This has profound implications for a region's well-being as a declining population and fewer working-age people reduce the size of the labor force as well as the number of people who contribute to pension and insurance schemes that support the older, retired population.

In the preceding chapters, we explored the economy and trade in detail, including the concept of comparative advantage, which requires regions to specialize and produce goods and service to which they are best suited (or, in economic parlance, produce those things with the lowest opportunity cost). Labor can confer a comparative advantage, though what that means for production depends considerably on the characteristics of the population. Demography is the study of populations and the relationship of population characteristics to important social phenomena such as economic productivity. Age structure can tell us a lot about the potential **labor pool**, or the number of available workers for the economy. **Educational attainment** data (e.g., literacy rate, percentage with a college degree, percentage graduating in different majors) provide insight into labor skill. Demographers use population **age structure**, or the number of people in different age ranges, to assess the viability of social security programs for the elderly, as these programs usually require large contributions from younger generations. Demographic data is the foundation for any comprehensive analysis of a population **cohort**, or group, such as Gen X, Y, or Z (and now, Generation Alpha, those born in 2010 or later). Given the importance of demographic data in both **cross–sectional**, or point-in-time, and **longitudinal**, or long-term, analyses, governments throughout the world invest significant resources in collecting population-level data. The largest of these undertakings is the population Census, which occurs once every 10 years in the United States. In 1790, the United States instituted its first decennial Census, the first country-wide population census, to count and characterize all persons and households in the country (1). The first section of this chapter discusses the questions asked in the short- and long-form versions of the decennial Census and the American Community Survey (ACS), implemented in the intervening years between consecutive decennial Censuses.

In the United States, demographic data are the cornerstone of a city's general plan, particularly for the Regional Housing Needs Assessment (RHNA) in California or equivalent in another state. As most land use and related decisions in the United States are regulated at the city or municipal level, the general plan and its various elements are a vital part of the American governance structure. The general plan helps identify local issues within the context of planning areas, such as transportation or housing, and population groups that bear disproportionate burdens in these planning areas. General plans are an important tool in addressing all three Es (environment, equity, and economy) of sustainability. In the United States, demographic data on births, deaths, marriages, and divorces, called vital statistics, is also collected by states and reported to the National Center for Health Statistics (NCHS) (1).

The collection of quality demographic data, or data that is collected from reliable methodologies implemented at regular time intervals and includes the entire population (i.e. the Census) or a generalizable sample, can help policymakers, businesses, and others describe populations, how they will change, and their likely impact on all aspects of society. Recently, for example, the Population Reference Bureau (PRB) released a report describing how US women in the 24-34 cohort today are in some ways worse off with respect to important health indicators than previous cohorts (This is an example of a health inequality or disparity). In addition to a population Census, the United States and most other countries conduct regular surveys on employment, health, business, social attitudes, and other topics used by policymakers, academics, businesses, and others. Some of these surveys, like the population Census, occur at regular intervals and are often undertaken or commissioned by some level of government; others occur more infrequently and are sometimes funded by the private sector. Data collection and **weighting**, or statistically adjusting, the **sample** data collected from a smaller pool of people to achieve some level of **generalizability** to a **population**, or larger group, of interest has become more complex with the transition to cell phone– and internet-based surveys. The errors in the 2016 and, to a lesser extent, 2020 U.S. presidential elections attest to the need for better data collection and weighting.

This chapter provides an overview of demographic concepts necessary for characterizing the modern world. Though many different demographic concepts and ideas and their relationships to each other are described in the chapter, all are concerned with either population composition or change. **Population** can refer to any cohort, or group of people, identified on the basis of a shared characteristic, such as all who bought a lottery ticket last week or were infected with COVID last year. However, in this chapter we primarily focus on populations in a particular geography. All concepts in the chapter can be used to describe people in a particular place and time. In the first section, we look at how demographic data is collected in the United States. Subsequent sections look at the principal components of population change (fertility, mortality, and immigration), age structure and dependency ratios, the challenges related to urbanization, and the relationship between demography and sustainability.

The U.S. Census

The biggest demographic survey in the United States is the decennial Census. The 2020 Census included questions on household size; age, gender, race, and ethnicity of household residents; **tenure** (i.e., whether the home is owned or rented); and relationship between other household residents and the **householder**, or the person in whose name the home is owned or rented (this person is sometimes called the head of the household). In the intervening years, the U.S. Census collects demographic information using a cluster sample of households throughout the country. This survey, called the American Community Survey (ACS), asks similar questions to the decennial Census to maintain comparability between data sets and adds questions on several social, housing, economic, and demographic characteristics (e.g., disability, fertility, plumbing, rent, home value, veteran status, employment status, occupation, etc.).

The race and ethnicity questions use a two-question format, the first of which asks respondents to identify if they are Hispanic or Latino. The second questions ask respondents to pick from a minimum of four race categories (White, American Indian or Alaska Native, Black or African American, and Native Hawaiian or other Pacific Islander) (25). The write-in areas and examples offered by the Census encourage respondents to provide more details about their race/ethnicity. This allows respondents to self-identify as they best see fit and the Census to develop details tabulations of respondent ancestry (e.g., German, Brazilian, etc.) (25).

Important Population Characteristics

Population projections are essential for planning purposes. Accurate predictions of the number of people and their characteristics can help the public and private sector anticipate and address future needs and issues. Population growth or shrinkage has significant impacts on a myriad of social, economic, political, and environmental factors. Businesses, for example, are interested in determining the number of likely consumers for their products. Demographic characteristics such as age, income, and education can provide valuable insight into a group's spending capacity and types of goods and services that are appropriate for the group. For governments, demographic characteristics such as poverty status, family households with children, or homeownership provide data on likely sources of tax revenues and the various types of needs in a community.

The components of population change are births, deaths, and migration. Demographers devote a lot of resources studying past and current trends to predict future birth and death patterns. Of the three components of population change, births are the most difficult to predict as the decision to have a baby is impacted by a variety of psychological, social, political, and economic factors. Births are also impacted by **fecundity**, or the biological capacity for reproduction. Deaths are comparatively easier to predict, though psychological, social, economic, and political factors play an important role, as do changes in medical and public health technologies. **Life expectancy** at birth, or the additional years babies can expect to live, has increased across the planet due to improvements in diet, sanitation, vaccinations, and medical care. Collectively these improvements in public health have reduced **mortality**, or death, from all causes. The result has been extended life spans and improvements in quality of life across all ages. In the United States and other developed countries, the elderly comprise an increasingly larger proportion of the population. By 2030, one in five Americans will be over 65; by 2060, there will be 19 million 85-year-olds as well as half a million centenarians (100 or over) (5). Both the public and private sector collect data across many domains (e.g., health, income, psychological, etc.) to better understand and serve the elderly, an increasingly important demographic group in the United States and many other countries. The elderly use health care and social services at a greater rate than those of working age. Shifts in age structure driven by increasing life expectancy and a lower fertility create a population pyramid that is increasingly top heavy. In the long run, an older age structure with increasing old age dependents is unsustainable and poses a serious threat to any country that is unable to reverse fertility declines or increase immigration.

Migration patterns within countries are more difficult to understand than those between countries as these are often dictated by national immigration policy. China, for example, has historically regulated rural to urban migration, whereas in the United States, people are free to move about and often do depending on economic, social, and political factors. For example, in recent years, individuals and businesses have moved out of California, citing high housing costs, increasing crime, congestion, taxation, and regulation. Retirement in the United States often provides a big impetus for moving, especially out of high cost-of-living areas, as income reduces significantly for many elderly households upon retirement. Weather can also play an important role as seniors sometimes move from colder northern locales to warmer southern states during the winter (this is sometimes referred to as the "snowbird" phenomenon).

Demographers often study **cohorts,** or groups of people who share some type of common demographic experience in a certain time period. Cohorts also can be identified as such because of this shared experience at some later time period (1). **Age cohorts**, sometimes called generations, are often referred to in both popular media and academia when discussing a variety of social, economic, and other phenomena such as attitudes toward social issues like gay marriage, support for social movements such as Black Lives Matter (BLM), job-seeking behavior, or climate change. Age cohorts often differ greatly with respect to important demographic characteristics such as life expectancy, marriage, education, and fertility. Other cohort characteristics provide important clues about a region's *zeitgeist,* the general intellectual, moral, and cultural climate of an era, and likely changes given the differences in younger age cohorts.

As mentioned earlier, public and private entities collect data on a staggering number of demographic characteristics. Demographic data can be expressed in terms of counts, rates, or ratios. Counts are absolute numbers. The population of a country in a certain year is an example of a count. Rates are the result of the division of two counts and are sometimes reported in terms of a percentage. Examples include the **crude birth rate**, which is the number of babies born in a year in a particular region divided by the total population of the region. The birth rate is usually higher in poorer countries and tends to fall as incomes rise. In 2020, the birth rate for high-income countries was 10 per 1,000 whereas the least developed countries were more than three times higher at 33 per 1,000 (6). Another important rate or indicator of a country's development level is the **infant mortality rate**, which is obtained by dividing the number of infants or children under 1 who have died by the total number of births. The infant mortality rate is also nearly five times higher (48 per 1,000) in the least developed countries when compared to high-income countries (10 per 1,000) (6).

Rates can also be expressed as ratios or can be divided by other rates to create a ratio. **Gender or sex ratios** are estimated by dividing the total number of males by the number of females and multiplying by 100 or 1,000. This provides an estimate of the number of males per 100 or 1,000 females. In some countries, female feticide—selective aborting of female fetuses—has resulted in highly skewed sex ratios favoring male children. A larger number of males relative to females impacts **nuptiality**, or the marriage rate, and birth rate as some men will be unable to find partners given the relative death of women; there is also some evidence for higher rates of social deviance such as crime, but the findings are not conclusive. Nuptiality is also concerned with the characteristics of those who get married, such as median age at first marriage, and the dissolution of marriages or divorce. **Household formation**, or the number of new households formed in a time period, is sometimes linked to marriage as younger, unmarried couples may not cohabitate and continue residing with their parents or guardians. Economic cycles, specifically downturns, have a profound impact on nuptiality and household formation. The 2007–2008 recession slashed household formation. About 1.5 million new households were formed annually between 1997–2007. In the 3 years after the recession, the number of new households formed per year fell to only 500,000 (7). Household formation impacts various aspects of society, such as demand for new housing, school enrollment, traffic, civic participation, and consumer spending. The number of new households formed increases as young people move out to live on their own, young couples move out to start their own households, and when marriages dissolve.

A related demographic concept is **headship rate**, or the probability that a person in a certain age range is likely to be head of household. In the U.S., headship rates for those between 18 and 34 fell from 37.9% to 35.8% between 2007–2011 (this was a 2.1 percentage point decrease or 5.5% decrease from the 2007 baseline). This decline was significant because nearly three fourths of the decrease in households formed during this time period were due to falling headship among those between 18 and 34. The recession created immense economic uncertainty among younger adults, who could not afford to leave home to start their own households. In the U.S., headship rates never recovered fully after the recession and were further pushed downward by the pandemic (8). Falling headship rates in the long term reduce demand for housing, which impacts the economy, reduces wealth accumulation by those living with parents or guardians, and increases inequality (8). Furthermore, householders, or heads of households, especially homeowners, tend to be more invested and involved in their communities. Civic participation also suffers with a decline in headship rates. There are some silver linings, however, as young adults living with parents spend more on luxury goods (31).

Researchers try to differentiate between the different factors that impact household formation. In some cases, demographic factors such as aging, migration, or those related to nuptiality play an important role. In others, economic factors such as a recession or a robust job market may increase or decrease household formation. By understanding and predicting headship and household formation trends, policy makers and planners can anticipate and address needs across different societal domains such as housing, transportation, education, and consumption.

Births (Fertility), Deaths (Mortality), and Moving (Immigration)

As discussed earlier, births and deaths, or fertility and mortality, are the fundamental drivers of population change. If births and deaths are in equilibrium, or roughly the same, population growth is slow, negligible, or slightly negative. For much of human history, both fertility and mortality rates were high and the populations stagnant (see Figure 6.1). Populations have also been reduced throughout history by various diseases, some of which have caused such widespread sickness and death that human populations did not recover from until decades later.

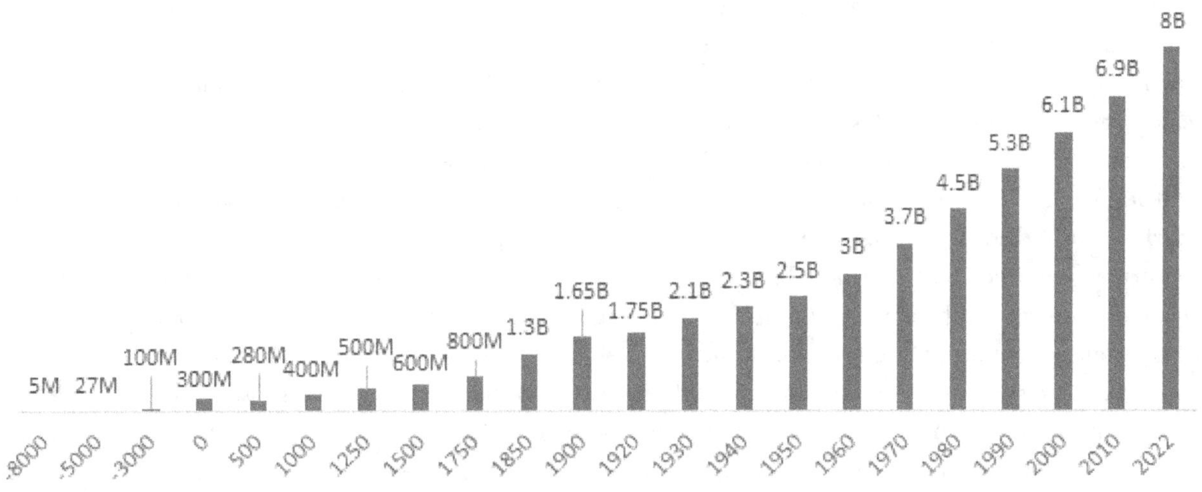

Figure 6.1   Estimated world population increase, 8000 BCE to 2022.

Adapted from United Nations and Population Reference Bureau Population Estimates, https://population.un.org/wpp/Publications/Files/Key_Findings_WPP_2015.pdf.

Bacteria and viruses, the two principal causes of disease in humans, predate *Homo Sapiens*. However, their ability to cause mass death was limited by the low population densities associated with hunter-gatherer and nomadic lifestyles. Urbanization brought many benefits; however, it also provided the densities necessary for epidemics and, in some cases, pandemics as diseases spread through traders, migrants, and travelers.

The bubonic plague likely first appeared in 541 AD in the Byzantine Empire and wiped out nearly a quarter of the human population in the next 2 years. It reappeared 800 years later in 1350 and killed nearly a third of the human race (10). The plague would reappear in the mid 1850s in China, killing 15 million people (10). In addition to the plague, cholera, measles, smallpox, and the flu have all caused epidemics and pandemics of various proportions and culled vast swaths of the human race. For example, Native American population declines after European colonization were partly due to diseases such as smallpox and measles for which they had no prior immunity (11). Some scholars theorize that colonization disrupted food supplies, increasing Native American susceptibility to diseases.

Medical science and public health advanced greatly in the 19th and 20th centuries with the development of germ theory, vaccinations, antibiotics, and sanitation. However, these could not stop the 1918 Spanish flu from claiming 50 million lives. The 2020 COVID-19 pandemic would have exacted an even heavier toll than the 6–7 million who died through mid 2023—and this may be an undercount—had it not been for the rapid and coordinated containment measures taken by governments across the world and rapid development and deployment of vaccines (32).

Humans only recently developed the capacity to significantly reduce the **morbidity**, or sickness, and **mortality**, or death, associated with highly infectious diseases. Population growth has been kept in check by diseases for much of human history. It was only in the last 2 centuries that medical science and public health reduced mortality from all causes. Additionally, innovations such as the Haber–Bosch process, which increased nitrogen availability for fertilizer and genetically modified crops, increased agricultural yields to provide sufficient food to allow for population growth. As shown in Figure 6.1, human population did not grow significantly till the mid 1800s; during the 20th century, it took roughly two decades to add each additional billion (32). Together these factors led to the largest increase in human population in history. Industrialization in the 19th and 20th centuries provided the burgeoning populations with employment and many consumer products that substantially increased quality of life for the masses. The increasing wages allowed for the growth of the middle class. Today, there are not only more humans on the planet, but they are also living longer and more comfortably than ever before. [1]

Life expectancy has increased considerably during the past 2 centuries due to improvements in public health, medical care, and standard of living. As recently as the 1800s, life expectancy was no more than 40 years, even in richer countries (1). Today, life expectancy at birth exceeds 80 in some countries and is 60 in the least developed countries (6). In the coming years, many countries will be home to large numbers of **octogenarians (80+), nonagenarians (90+), and centenarians (100+)**. The increase in life expectancy coupled with the decline in fertility has transformed age structure. Throughout history the age structure has tended to resemble a pyramid—hence the term *population pyramid*—with many more children and younger adults than older adults. In many countries with longer life expectancies and fewer births, the age structure has become somewhat inverted with many more working-age people and elderly than youth and children. If low fertility persists, even countries with substantial middle-aged populations will become top heavy over time as the middle-aged population becomes older with fewer younger generations to support them by paying into Social Security, pension, and insurance schemes. In the United States, the most recent report by the SSA on the solvency of the Social Security program was optimistic about the U.S. fertility rate, assuming it will return to 2.0. The fertility rate was another casualty of the 2007–2008 recession, falling from replacement level (2.1) in 2005 to 1.9 in 2010 and 1.7 in 2019. The relative prosperity of the past 10 years, from 2010 to 2020, has not increased the fertility rate. The United States has experienced fertility rebounds in the past as the fertility rate fell from a high of 3.6 during the 1960s to 1.8 in the mid-1970s and later rebounded to 2.1. However, the optimism of the SSA may be unfounded as the reasons for the recent fertility declines are complex and different than the past (12).

---

1. This perspective, of course, is a hagiography of modernity as it elides the major wars and sustainability issues that accompanied our march into the 20th century and beyond.

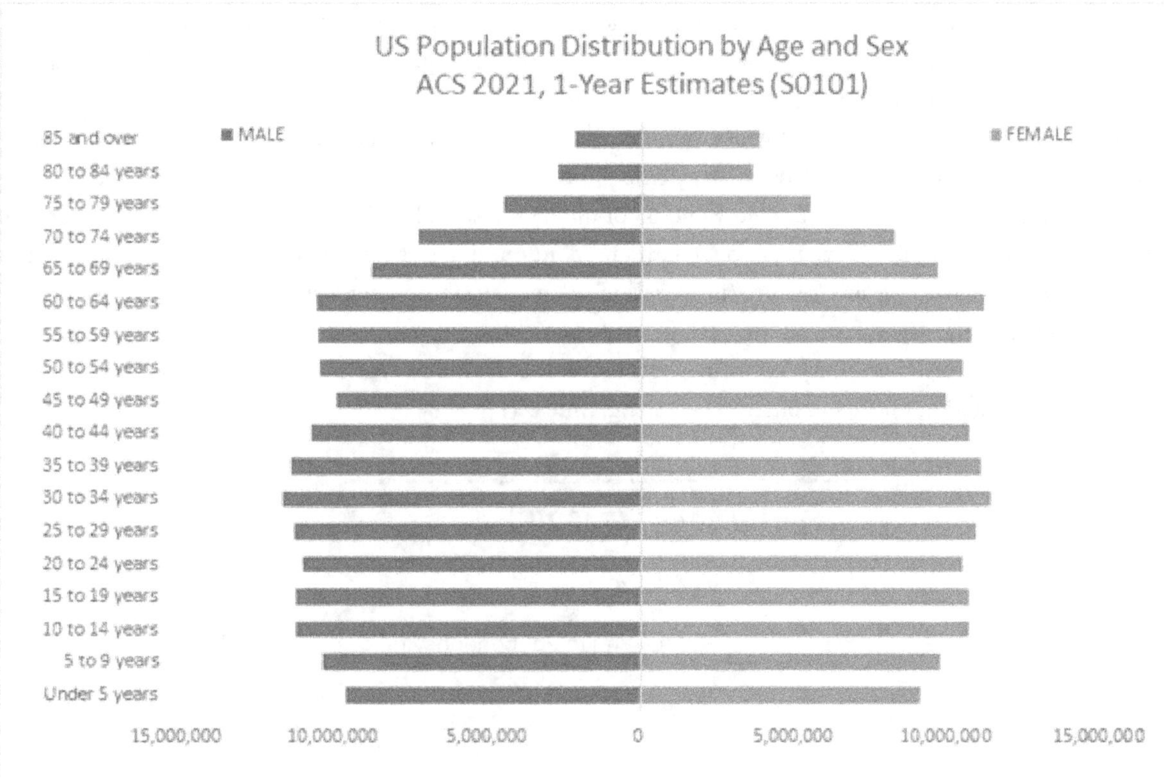

Figure 6.2   **U.S. population distribution by age and sex.**

Adapted from ACS 2021, 1-year estimates (S0101), https://data.census.gov/cedsci/table?q=1-year%20age%20and%20sex&tid=ACSST1Y2021.S0101.

The number of births in a region, as discussed earlier, is the result of many different social and biological factors. Social factors can be economic, cultural, psychological, or legal. For example, the Supreme Court of the United States recently ended federal protection for abortion by overturning a nearly 50-year decision, *Roe v. Wade*. The likely impact of this decision will be a full or partial ban on abortions in many states. This will likely increase fertility in those states. The 2007–2008 recession and the pandemic both impacted the labor market. Younger adults were particularly hard hit, and many chose to delay marriage and childbirth. The fertility rate tends to be high in countries where women don't exercise full control over their reproductive health decisions. Adolescent fertility is also high in many of these same countries: In 2021, over 13 million babies were born to mothers under 20 years of age (3). In general, only a few regions of the planet currently have fertility rates that exceed **replacement level (2.1)**, the number of babies per woman at which the population replaces itself: sub-Saharan Africa (4.6); Oceania, excluding Australia and New Zealand (3.1); Northern Africa and Western Asia (2.8); and Central and South Asia (2.3). In countries with below replacement level fertility, populations will decline over time.

Declining births will reduce populations and change the age structure in favor of the middle aged and elderly. The population pyramid will slowly morph into a vase with a smaller base, large middle, and thinner neck. As discussed earlier, this has significant repercussions for society as younger people pay into Social Security, pension, and insurance schemes, which are disproportionately used by the elderly. An attenuation of the working-age population causes these schemes to become insolvent more quickly. A drop in fertility, however, is not all bad, especially for developing countries with high fertility rates and few elderly due to lower life expectancies. Fertility reductions can lead to a **demographic dividend**, an age structure in which the working-age population exceeds the nonworking age (those below 16 and those above 65). A demographic dividend can spur economic development as the potential labor pool is large and relatively less burdened with expenditures for the nonworking age (3). The rapid ascent of South Korea, for example, is partially attributed to their demographic dividend (13). This ratio of **working age** (15 to 64) to nonworking age or **dependents** (14 or below and 65 or older) is also called the **dependency ratio**. If dependency is driven by large numbers of elderly and fewer youth and children due to falling fertility, the long-term prospects for the economy can be negative and require immigration, increase in retirement age, and increases in productivity across all ages. Europe and Japan have an increasing number of old-age dependents, and other countries at or above replacement fertility are also declining. Reversing fertility can be difficult. Some have managed to incentivize children through a combination of tax and other benefits (e.g. time off).

$$\frac{(\#0\text{-}14) + (\#65 \text{ or more})}{\#15 \text{-} 64} \times 100$$

Figure 6.3   Calculating the dependency ratio.

Fertility

Fertility can be measured several ways. All measures are based on the number of live births in a particular area in a certain time period, usually a year, and the population, or a subset of the population (e.g., women of reproductive age) of that area during the time period of interest (1). The **crude birth rate (CBR)** is calculated by dividing the number of births in a year by the total population. In an area with 5,000 births and 100,000 people, the CBR would be .05 (5000/100,000). Often, demographic data is reported in terms of 1,000 or 100,000 people, so a CBR of .05 would be 50 (.05 x 1,000) per 1,000. The CBR, like other crude measures, can be calculated for different population subgroups based on age, race/ethnicity, or geography. The general fertility rate (GFR) is calculated by dividing the number of births by the number of women between 15 and 44, the prime reproductive age range. The age-specific birth rate (ASBR) is calculated by dividing the number of births in every age (10, 11, . . . , 42, 43, 44) by the total amount of women in that age. The total fertility rate (TFR) is simply the sum of all ASBRs. The examples in Table 6.1 illustrate the relationship between ASBR and TFR.

Table 6.1. 2020 Births, ASBR, and Number of Females by Age of Mother

A	B	C	D	E
Age	**Babies**	**Number of Females**	**ASBR**	**ASBR x 5**
10 to 14	1765	8825000	0.0002	0.001
15 to 19	157548	10237255	0.0153	0.0765
20 to 24	663732	10568981	0.0628	0.314
25 to 29	1022033	11355922	0.09	0.45
30 to 34	1067798	11263692	0.0948	0.474
35 to 39	562833	10886518	0.0517	0.2585
40 to 44	120278	10193051	0.0118	0.059
45 to 54	9214	10237778	0.0009	0.0045
			TFR	**1.633**

Adapted from "Births: Provisional Data for 2020," NVSS, Vital Statistics Rapid Release (15).

The TFR can be calculated using the data in Table 6.1. Column A shows the age ranges of the mother, column B provides the number of live births, and column C indicates the number of females in the age ranges. Column D estimates the age-specific birth rate by dividing column B by column C, and column E multiplies the data in column D by 5 as there are five ages in each age range. The sum of column E is the TFR for 2020.

Age	ASBR/1000			ASBR/1000 x 5		
	2000	2010	2018	2000	2010	2018
10 to 14	0.9	0.4	0.2	0.0045	0.002	0.001
15 to 19	47.7	34.2	17.4	0.2385	0.171	0.087
20 to 24	109.7	90	68	0.5485	0.45	0.34
25 to 29	113.5	108.3	95.3	0.5675	0.5415	0.4765
30 to 34	91.2	96.5	99.7	0.456	0.4825	0.4985
35 to 39	39.7	45.9	52.6	0.1985	0.2295	0.263
40 to 44	8	10.2	11.8	0.04	0.051	0.059
45 to 49	0.5	0.7	0.9	0.0025	0.0035	0.0045
			TFR	**2.056**	**1.931**	**1.7295**

Adapted from National Center for Health Statistics, online data portal (16)

The first three columns of Table 6.2 show the ASBR per 1,000 women. The next three columns take the figures in the first three and divide them by 1,000 to obtain the crude age-specific birth rate and then multiply the result by 5 as there are five ages in each age range. The TFR is calculated by summing the data in each of the last three columns.

The trends in U.S. fertility in the last 2 decades are clear. The TFR in 2000 was 2.06, and the current TFR is closer to 1.6. As discussed earlier, some policymakers have argued that U.S. fertility will rebound. Some call this perspective Panglossian, or excessively optimistic, as the decline in overall fertility is driven by declining ASBR among women younger than 30. Conversely, as shown in Table 6.2, the ASBR for women over 30 rose between 2000 and 2018 and only slightly declined in 2020. This indicates that many women are delaying childbirth until their 30s or later. This choice to delay childbirth is driven by social and economic factors such as the 2007–2008 recession, greater female labor force participation, and changing cultural norms regarding work, children, and marriage. As women choose to delay childbirth, fecundity, or the biological ability to have children, also drops due to aging. The **opportunity cost for early marriage** is much greater for women as childless women tend to have higher pay; men do not generally experience any negative wage or career impacts due to childbirth (17). Additionally, women who delay marriage tend to marry spouses with lower wages (17). The relative dearth of men at a comparable socioeconomic status may act as a deterrent for some women who have chosen to delay marriage. This likely has an impact on the decision to have children as cultural norms, though shifting, still encourage childbearing in a traditional two-parent household. Other factors may also be impacting fecundity worldwide. Some researchers speculate environmental stressors have reduced both sperm and egg quality throughout the world (9, 18). Urbanization and industrialization play a major role in increasing exposure to environmental stressors. In many developing countries, the deterioration in fecundity is likely smaller as women still have children when young. This advantage is likely to disappear as poorer countries develop economically and marriage and childbearing ages are pushed upward.

Mortality

All biological organisms seek to survive until they can pass on their genes. Humans, however, don't define the purpose of life exclusively in terms of reproduction. As falling fertility suggests, many choose not to reproduce. However, the instinct for survival remains, and even those who remain childless usually seek to live as long as they can. Modern medical care and public health measures such as sanitation have allowed modern humans to greatly exceed the life expectancy of their ancestors. Measuring death, when it occurs and how it occurs, is an important function of demography and public health.

Every living creature is "at risk" for death from accidents, predation, or illness. Humans also kill each other for a variety of reasons that have nothing to do with survival. Suicide appears to be a uniquely human phenomenon. Suicide rates have risen in certain regions in the aftermath of the 2007–2008 recession and among certain population cohorts. A more detailed discussion of suicide will be provided in Chapter 8. The most basic mortality measure is the **crude mortality rate**, which is estimated by dividing the number of deaths in a region in a specific time period by the population of the region (all rates are place and time specific). Like other rates, the result is often multiplied by 1,000 or 100,000 to provide an estimate of the rate per 1,000 or 100,000 persons. The crude death rate for the United States in 2020 can be estimated by dividing the total number of deaths (3,383,729) by the total population (329.477 million). The results of this calculation (3,383,729 / 329,477,000 = .01027) are multiplied by 100,000 to calculate the crude death statistic of 1,027 deaths per 100,000 persons. Crude death rates can be expressed in terms of a cohort defined by age, illness, or some other characteristic. **Age-specific death rates** are the number of persons in an age range who have died divided by the total number in that age range. **Disease-specific mortality rates** are similar except the numerator includes all who have died from a disease and the denominator is all who had the disease.

Mortality analyses are used to create **survival rates**, the probability of surviving from one age to the next. In the United States survival rates are 90% or higher until 65 (1). In other words, the probability of surviving from one year to the next is 90% or higher for those below 65. Survival rates are often presented in a life table and include proportion dying, number surviving, number dying, and person-years lived in the age interval, total person years to be lived, and **life expectancy**, the additional number of years a person at a particular age can be expected to live or the average number of years remaining for persons who have attained a certain age. For example, in 2018, the life expectancy for a 20-year-old in the United States was 59.5 (26). This means an average 20-year-old could expect nearly another 60 years of life. In the same year, an 80-year-old could expect to live 9 more years (26). Life table–type analyses can be applied to many other demographic phenomena such as marriage, employment, education, or housing stock (1). In each case, the life table shows the expected number and percentage dissolving every year. A marriage table, for example, would show the percentages of marriages surviving every year. These types of tables are extremely valuable for policymakers to project needs in different areas such as medical care or housing. Insurers use life tables to price and sell life insurance. Similar methodologies are used for pricing and selling different types of insurance (e.g., automobile, health) as all insurance schemes rely on probability of usage of certain services and likely costs of the services. By estimating the likelihood of usage and cost, insurers can spread the cost out to all those who are insured with the understanding that only a subset, or few, will use the insurance.

A cause of death is sometimes difficult to assess as an individual may present with multiple **comorbidities**, or ailments, before dying. A death certificate requires only one cause of death. In some cases, the official certifying the death may not be familiar with the person's medical history and may indicate the wrong cause of death, citing only proximal or immediate cause and not the other causes or **underlying cause**, the disease or condition that initiated the events leading to death. The CDC "Cause-of-Death Certificate" provides medical officials four spaces to provide a "chain of causes" proceeding from the underlying cause of death to the immediate cause of death. For example, a ruptured heart muscle (immediate cause) is the result of a heart attack caused by a blood clot, which was the result of heart disease (underlying or primary cause) (27). The complexities in attributing cause of death were on full display when some argued that COVID deaths were overrepresented as COVID was not the primary cause of death for patients with other underlying conditions or comorbidities. However, while the mortality (and morbidity) risk to those with underlying conditions was higher, it is unlikely they would have prematurely died in the absence of a COVID infection.

Migration

Migration is an important component of population change and the factor easiest to control through policy. Governments usually control national migration through some type of formal application process for work, study, residency, or citizenship. Many countries allow a pathway to citizenship for those willing to invest significant sums of money in the country. The United States, for example, allows a pathway to obtain a green card, or lawful permanent residence, and then citizenship, for $1.05 million invested in the United States through the EB-5 Immigrant Investor program. Though there are many ways to reside, study, or work in the United States, only **green card holders**, those who have been granted permanent lawful residency, may eventually apply to become naturalized citizens. Migration within countries is more difficult to predict, especially if internal migration is unrestricted, as it is in the United States. People move for a variety of reasons, including work, study, climate, marriage, or retirement. In some cases, migration is permanent, and in others, it is temporary or seasonal. Economic factors are often important and include job markets, onerous taxes, or high cost of living. Some countries regulate internal migration through a visa process. In China, for example, the Hukou (household registration) system classifies people through their place of residence, usually based on place of birth, which confers eligibility for government benefits (29). The system was created to regulate rural to urban migration as rural workers could go to cities for temporary or informal work but did not have access to the same benefits as those with a city hukou. Obtaining a city hukou is contingent upon demonstrating job stability and payment into the local government insurance schemes for a certain period of time (29).

Any measure of migration will involve making determinations about the characteristics of an **immigrant**, someone coming into an area, or **emigrant**, someone leaving the area. These include factors such as place of usual residence, minimum distance required for establishing migration, and time intervals required to differentiate between temporary and permanent moves. Whatever the criteria used, migration is an important component of population change as both temporary and permanent migrants alter the social fabric of the places they are moving to by their actions in a variety of domains (e.g., economic, education, recreation, marriage, etc.), especially their birthing decisions; in the United States, migrants often have higher fertility rates, which tend to fall with subsequent generations whose marriage, birthing, and other decisions mirror those of other native-born Americans. The places emigrants leave behind are also altered profoundly by the loss of labor, consumption, social and cultural actions, and taxes. Skilled migrants leaving developing countries can constitute a "brain drain" as the flight of human capital significantly dampens economic development in their native countries. However, these same migrants often send significant sums of money back to their native countries as **remittances**, which positively impact local economies. Researchers have also identified a "beneficial brain drain" (BBD) effect as the prospects of emigration can foster greater investments in education in the native country raising average education levels. Simply put, emigration provides job markets for the educated and skilled in a poor country, who would otherwise lack the incentives to acquire education or skills, and since only a portion of the educated or skilled are able to leave, the average level of education in a poor country will be higher, as many educated would remain, than it would be in the absence of emigration (33).

Urbanization

More than half of the planet's population currently resides in an urban area. This trend will continue in the coming decades until over three quarters of the planet's population lives in a city. Urbanization and globalization reinforce each other. Urbanization provides labor, clusters of industries, economies of scale, and markets. Globalization provides a growing economy, with cities at the epicenter. Globalization also reshapes the cultural fabric of cities by introducing new foods, languages, art, music, ideas, and people. The economic and cultural transformations in the city exert a pull that drives rural to urban migration. Many come looking for better jobs and new opportunities in different life domains: Cities have a myriad of subcultures, recreational activities, social venues, and people for friendship and romance. Today, over 80% of the world's GDP is generated by cities (20, 23). North America (82%), Latin America (81%), Europe (74%), and Oceania (68%) are the most heavily urbanized. About half of Asia is also urbanized. Sub-Saharan Africa continues to be more rural (57%), though this will change in the coming decades (20).

Demographic transition theories can help shed additional light on the growth of urban areas. Demographers divide demographic transition into pre-industrial, industrial, and postindustrial phases. During the pre-industrial phase, birth and death rates are high, so there is little population growth. People living in urban areas are likely to have higher mortality rates than those in rural areas due to poor sanitation, high population density, and greater exposure to environmental contaminants. Those in urban areas also have lower fertility rates, so the urban population is only replenished by rural migration. As urban areas develop, mortality drops as cities add needed infrastructure including housing, reduce pollutants, and improve sanitation. Life expectancy improves in urban areas and migration continues, adding to the urban population. The drop in mortality, coupled with higher fertility, particularly in rural areas, creates a large **natural increase**, or growth driven by excess births, of the overall population. The urban population swells through both a natural increase and migration. Eventually, as more of the country becomes industrialized and urbanized, mortality and fertility continue to drop with a concomitant increase in life span. The dependency ratio, or ratio of working age to nonworking age, starts getting smaller as there are fewer younger people to replenish the working and a larger elderly population. Some countries in this postindustrial stage experience slow growth or negative growth. Japan is a notable example, but there are many others, including China, facing demographic contractionary pressures. Demographic forces are a powerful complement to economic theories for understanding the rise and fall of regions and nations.

There is no historical precedence for the kind of urban transformation occurring today. Many of the rapidly developing urban areas today are not mega-cities but regions with less than 500,000 people (19). Small and midsize cities will proliferate, particularly in Asia and Africa: Between 2018 and 2050, more than 90% of the urban growth is expected to occur in these two regions (20). Asian Cities with million or tens of millions are not expected to significantly increase their share of the growing urban population. By the century's end, the world's population center will have decisively shifted to Asia and Africa, with the majority living in smaller and newer urban areas.

Urbanization is not a uniform process; however, irrespective of how cities form and grow, all urban areas must provide essentials such as food, water, sanitation, jobs, housing, health care access, and other infrastructure. Cities must be safe and provide an environment in which all ages, from the young to the elderly, can thrive. The demands on the city are considerable, and many are unable to meet the challenge, especially when rural to urban migration is high. Lack of adequate and affordable housing is an ongoing challenge in many countries. In developing countries, many of the urban poor often live in slums. In some regions, more than half of the urban population lives in slums, and some of the largest slums have hundreds of thousands or more residents. In 2018, roughly a quarter of the urban population worldwide lived in slums. The countries with the highest proportion of slum dwellers were in sub-Saharan Africa (56%). In countries like the Central African Republic (98%) and Sudan (94%), nearly all urban residents lived in slums (21). In Eastern Europe, some countries (e.g., Hungary, Romania, Ukraine) have between 10%–15% of urban residents living in slums. In Latin America and the Caribbean, about one in five urban residents live in slums (21). India has the largest number of slum dwellers, with over 65 million people living in slums. India's largest slum, Dharavi in Mumbai, is home to around a million people. In the United States, California has the largest share of the homeless population, with many living in makeshift tent cities. These are not slums on the scale found in Asia or Africa, but they share many of the same characteristics.

Slum dwellings are usually of poor quality and durability that are often built in hazardous sites and are unable to provide protection against climactic conditions such as rain, health, cold, or humidity. Crowding is common as living areas are insufficient. Sanitation is inadequate as homes often lack bathrooms with plumbing. The absence of adequate sanitation results in high rates of communicable disease. Diarrheal diseases are common and are a major source of morbidity, or sickness, especially among children, and lead to many missed school days (24). Water infrastructure is often inadequate as many homes don't have plumbing or access to a protected drinking water source, free from contamination. Water-based diseases such as cholera and typhoid fever can be rampant, increasing both morbidity and mortality. Many who live in slums are employed as day laborers and lack typical work-based benefits such as sick days, vacation, disability or health insurance. Sickness exacts a heavy economic toll on a population that is already at the very bottom of the income distribution. The relationship between public health, work, and economic development will be further explored in Chapter 8. Slums dwellers often lack legal renter or homeowner protections, or **security of tenure**, making them vulnerable to evictions or other harassment and punitive actions by their landlord, government, or other entities (22). Slum dwellers face additional risks as they are more exposed to a variety of environmental hazards and often live in areas that are more susceptible to natural disasters and the impacts of climate change (22).

Urbanization and Sustainability

The growth of urban areas impacts all dimensions of sustainability. Urban poverty, which will be discussed in the next chapter, is a feature of all urban areas throughout the world. Income and wealth inequalities are rising everywhere. The spoils of economic growth, much of which is propelled by urban areas, is not shared equally by all. Mismanaged urban growth can also increase pollution, diseases, and resources consumption. Suburban sprawl, as discussed in Chapter 2, is more resource intensive than higher density development. Additionally, sprawl requires greater investment in automobiles and related infrastructure, leading to more GHG (green house gas) emissions. In the United

States, sprawl ,coupled with the front-yard setback mandated by single-family zoning, resulted in a front-yard water-intensive lawn. Only recently, due to droughts in some areas, has the green front lawn been replaced by native plants, or other nongrass landscaping. City dwellers tend to have a higher carbon footprint as their greater income, relative to rural residents, allows them to consume and waste more. Cities generate more than three quarters of the CO_2 emissions and consume more than two thirds of the world's energy (20). The environmental impacts are less when adjusted for income; however, this means that as the world develops and becomes more urbanized, consumption and waste patterns of previously poorer countries will mirror those of their richer counterparts. Another major impact of urbanization is the loss of natural habitat. Cities grow at the cost of the natural environment: Biodiversity loss increases as natural habitats and ecologies are razed to make way for new developments when cities grow to accommodate those who flock to them in search of the same lifestyle as others who are benefitting from globalization.

Conclusion

Auguste Comte, the father of positivism, is said to have remarked, "Demography is destiny." The implications of this statement are being felt in many different ways today. Populations face existential threats from outside and, in the case of very low fertility, within. Populations change in response to a myriad of forces, some natural (epidemics, climate) and others social (economic cycles, legislation, cultural mores). Populations also shape these factors. Some will survive and adapt, while others will perish. Some sound the alarm of too many people, and others, such as Elon Musk, argue that fertility declines are the most dangerous trend facing humanity.

Urbanization, globalization, and environmental pressures impact morbidity and mortality in a myriad of ways and shape choices made by people regarding births and migration. These choices, in turn, shape how urbanization and globalization unfold and the impacts on the environment. As populations grow, shrink, and evolve in different ways, governments often try to steer them to desirable ends. Some do it by force, others use incentives like tax breaks, some appeal to the population's good will, and others seek to educate. Many simply don't do anything, even when some intervention is warranted. Businesses, community leaders, and others also try to exert influence on people and their demographic choices through various cultural mediums such as religion, entertainment, or social media.

The recent reversal of *Roe v. Wade* is just one example of how laws impact birth rates and demographic trajectories. China's former one-child policy provides another example of using laws to shape demography to achieve a desirable social goal. The policy reduced TFR to 1.0, allowing China to control population growth and focus on economic development. The government ended the policy in 2016 to reverse increasing old-age dependency as rising prosperity increased life expectancy but fewer future workers were being added due to low fertility. The Chinese TFR, however, has hovered around historical norms (1.0) despite an end to the policy. Low fertility represents perhaps the gravest threat to China's long-term prosperity. The same story is unfolding in many countries throughout the world. In general, declining fertility, coupled with longer life expectancy, does not bode well for pensions, insurance, and other similar schemes that depend on contributions by younger, working-age people. There are many other similar stories related to demography. Demographic choices are not set in stone and can change quickly. Some of the trends we see today may reverse themselves, or the trends may continue, and the human race will age and reproduce less. This will be beneficial for the environment, flora, and fauna, in many ways, but the future of humanity itself will be more precarious!

Chapter 7

Measuring Poverty, Economic Development Approaches, and Poverty Policy

Key Terms

Relative poverty: Poverty thresholds in developed countries are usually based on local or national income distributions such as area median income (AMI) or on predetermined income levels such as the U.S. federal poverty levels.

Absolute poverty: Poverty thresholds for extreme poverty. The first threshold used $1 per day as the extreme poverty line to measure the number of people worldwide who fell below the line. The thresholds have been updated several times since then to account for changes in purchasing power. The most recent update created a new line at $2.15 per day.

Structural adjustment policies: Policies and conditions countries must follow to receive loans from the IMF or World Bank. These conditions typically emphasize deregulation, free trade, reduction in government spending, and privatization. These policies have been criticized as being overly focused on debt repayment at the cost of other necessary investments in health, infrastructure, education, and development that also develop human capital and spur economic growth.

Other key terms will be in **bold** and defined in the text.

Review Questions

- What are the different ways to measure poverty in developed and developing regions?
- What are the impacts of COVID on poverty?
- What are the various causes of poverty?
- What are the various approaches to national economic development and the shortfalls of these approaches?
  - What is dependency theory?
  - What are structural adjustment policies?

- ○ What is import substitution industrialization?
- What is the EITC, and how does it work?
- What is UBI, and how does it work?
- What is total factor productivity and factor accumulation?
- What is the EITC, and how does it work?

Introduction

In 2020, COVID-19 brought the world's economy to a standstill. The economic impact was felt most acutely by the poor in both developed and developing countries. COVID's economic impact was greatest among the urban poor and likely exacerbated urban inequality (40). Extreme poverty also increased by around 1% or more. This means 70–115 million people fell below the international poverty line, which was recently revised from $1.90 to $2.15 per day (41, 42, 44). In 2021, the world's economy rebounded. In 2022 growth slowed to 3.5% from 5.7% in 2021, driven by inflation, the Ukraine war, rising interest rates, and other regional specific factors (43). In 2023, growth is expected to further slow to 3%. The economic growth in developing countries will be less than pre-COVID levels in 40% of developing countries, though some, like India, have rebounded considerably from -6% to 7% in 2020–2021 to 7%–8% in 2022–2023. Today, between 700–800 million people live in extreme poverty. Before COVID, this number was closer to 600 million (45).

Policies that aim to reduce poverty and related ill effects can also be controversial as they may create distortions in the labor market by disincentivizing work (1). To increase labor force participation among welfare recipients, the Clinton administration in 1996, through the Personal Responsibility and Work Opportunity Act (PRWORA), enacted reforms that reduced benefits, increased work requirements, and promoted two-parent families (2). Similarly, the generous benefits in some EU countries may create lifetime dependency as welfare benefits sometimes exceed the gross income from minimum wages in some countries. Additionally, the significant reduction in benefits when working for minimum, or near-minimum, wage results in a high marginal tax rate (as workers have to procure the lost benefits using earned income), which further disincentivizes work. In other words, when wages from work put workers above the poverty line or minimum thresholds for public assistance (health care, housing, food, etc.), they have to use the money they earned to purchase the benefits they have lost; in some cases, the additional wages may simply not be worth it, and it may make more sense for workers to not work or work less so that they stay below the poverty line.

More recently, some have attributed the worker shortages during the COVID-19 pandemic to generous government benefits enacted to offset the work disruption induced by the pandemic (4). This view is not without its detractors. Some argue that the quality of available jobs coupled with paltry pay and benefits are culpable for persistent unemployment among welfare recipients (5). Other suggest that many in poverty simply do not have access to the training and education required for higher paying jobs (6); additionally, the wealthier, as discussed in previous chapters, benefit disproportionally from rent-seeking behavior, which allows them to increase both income and wealth passively through investments without contributing to the economy.

These complexities notwithstanding, persistent poverty and inequality have profound multidimensional impacts on human well-being and society (7, 8). At the individual and household level, poverty impacts health and well-being directly and indirectly. The poor have less money to purchase essential and discretionary goods and invest in themselves, and they are more likely to suffer ill effects from wage disruptions as they have low savings and reduced access to insurance. Poverty impacts the procurement of quality health care, education, and housing, which impacts morbidity and mortality, earning potential, and income mobility and increases exposure to crime and violence.

In previous sections, we looked at the importance of free and fair markets to economic efficiency. Markets cannot function smoothly, allocating inputs or scare resources to their best use, if access to markets is restricted or if parts of the market are hidden or obfuscated so that buyers and sellers have incomplete or inaccurate information. Equality of access and transparency are not simply desirable social goals for markets but are essential for the long-term viability of markets and the economy. At a societal level, concentration of wealth and power can lead to macroeconomic instability as resources are misallocated by the wealthy as they pursue rent-seeking policies to protect their interests and maintain their positions. Social cohesion is compromised as many view the system as inherently "rigged." This can eventually lead to a crisis and fracture civil society. The recent upheavals in the United States are in no small measure due to rising economic inequality and how it intersects with race and ethnicity.

This chapter examines how poverty is measured in the United States and developing countries, historical approaches to economic development at the national level, and poverty policy in the United States. Poverty is a complex phenomenon as income poverty only captures one dimension of human development. At the household and individual level, public policy can mitigate the impact of income poverty through various programs and policies that provide essentials such as food, education, and health care. Poverty can also be addressed through economic development at all levels of society, from the local to the national. The World Bank, IMF and WTO, the institutions of global trade and finance, play an important role in economic development at the national level. Within nations, economic development and poverty policy is usually guided by national and local policy. In the United States, economic development often falls to states and cities, whereas the large antipoverty programs are often funded, directly or indirectly, through block grants (i.e. funding given to states and large metropolitan areas based on population size) by the federal government. Whatever approach is taken, ending extreme poverty and reducing wealth and income inequality are important Sustainable Development Goals (SDGs) and the goals of many policymakers throughout the world.

Measuring Poverty

Household income is usually used to determine poverty status. Households can be single person or multi person. If the total household income falls below the poverty threshold for that region, all members of the household are assumed to be poor. Data collection in poor countries can be difficult as some countries lack the capacity to survey households regularly. Between 2002 and 2012, there were 29 countries that did not collect any household economic data; others implemented surveys at large intervals (e.g., every 5 years), making it difficult to gauge current trends (23). The World Bank works with local statistical agencies in countries that need to build the capacity for data collection. In developed countries, such as the United States, data is collected regularly. In the United States, several economic and labor market surveys are carried out by the U.S. Census, Social Security Administration (SSA), Bureau of Labor Statistics, and many other public and private entities, including think tanks, academic institutions, and non-government organizations.

The cost-of-living differences between countries and within countries make it difficult to measure poverty. **Poverty thresholds**, or the income below which households are considered poor, are set by each country and are usually **relative** or based on the income distribution in the country. Poverty thresholds can also be **absolute**, a level below which households are considered poor, irrespective of country. The international poverty line, currently set at

$2.15 per day, is an example of an absolute threshold. This is the threshold being invoked during any discussion of **extreme poverty** in the world.

Often poverty thresholds are some percentage of the median income in the country. For example, in the European Union, households with income below 50%–60% of median income are usually considered poor (10). The federal poverty thresholds in the United States, however, do not use median household income but are instead based on work done by Mollie Orshansky, an economist at the SSA. Orshansky developed the thresholds in 1963–1964 based on the cost of an "economy food" plan for different household sizes and the observation that Americans generally spent about a third of their income on food (12). The original thresholds were derived by multiplying the cost of the economy food plan, which provides the minimum necessary calories for all members of the family, for different household sizes by three. For example, the 1963 threshold for a nonfarm family of four was $3,108. This translates to an annual economy food plan cost of $1,036 (or 3,108 divided by 3) (12). In other words, a nonfarm family of four in 1963 could feed itself for $1,036 and any family living below three times $1,036, or $3,108, would be considered poor. These thresholds have been updated annually for inflation using the consumer price index (CPI-U). The biggest criticism of the U.S. poverty thresholds is that they neglect variations in income and cost of living in different regions. As many federally funded poverty programs are tied to the poverty thresholds, many in higher cost of living areas are ineligible for government aid despite having inadequate incomes for basic needs like housing.

In 2020, the area median income (AMI) in the United States fell 2.9% to $67, 521 (11). **Area median income (AMI)** is often reported instead of average income as it's a better measure of the typical household income in a region (average incomes tend to be skewed by outliers or incomes that are really high or low). The number of people who reported earnings also decreased by 3 million (11). The decreases in working adults and earnings were the result of COVID-19 related shutdowns. The poverty rate rose from 10% to 11.4%, the first increase after 5 years of declines (11). This meant that roughly one in eight Americans was living in poverty during the early part of the pandemic. Educational attainment was correlated with income loss, with the largest income declines occurring among those without a high school (HS) diploma, who reported a median income of $29, 547, a 5.7% decrease from the previous year. The second most impacted group was those with a HS diploma and no college, who reported a median income of $47, 405, a 3.9% decrease from the previous year (11).

Area median income (AMI) is sometimes used as the benchmark for deciding various thresholds for middle income, or middle class, low income, and very low income. Using AMI allows policymakers to be regionally specific and account for regional differences in income and cost of living. The U.S. Department of Housing and Urban Development (HUD), for example, uses regional median family incomes (MFI) to decide the upper thresholds for low income (LI; 80% of FMI), very low income (VLI; 50% of MFI), and extremely low-income families (ELI; 30% of MFI). HUD also adjusts MFI for family size, with a family of four as the base case. The 2022 statewide median family income for California is $101,000 (26). Based on this, a family of four must make less than 80%, or $80,650 to be considered low income, less than $50,500 (50% of MFI) to be considered very low income, or less than $32,500 (30% of MFI) to be considered extremely low income (26). The adjustment for a family of two is 80%, so a family of two must make less than $64,550 (80% × 80% × 101,000) to be considered low income (25). HUD administers the funds for different housing assistance programs, including housing vouchers (Section 8), the biggest rental assistance program in the country. Eligibility for Section 8 and the 25 other housing programs are income based. A household paying more than 30% of its household income toward housing is considered cost burdened. The Section 8 program makes up the difference between fair market rent (FMR) for a region and what an LI, VLI, or ELI household can pay based on the assumption they can contribute no more than 30% of their income. The following example shows how a housing subsidy is estimated for a VLI family in California:

Estimation of Housing Voucher Subsidy for a Very Low Income Family in CA.

FMR = $2,000 annual FMR = $2,000 x 12 = $24,000

Monthly income required for FMR = $2,000 / 0.3 = $6,667 annual income = $6,667 x 12 = $80,000

(We divide by 0.3 because a household can pay no more than 30% of their income toward housing.)

FMI in CA = $101,000

VLI family in CA = $101,000 x 0.5 = $50,500

(We multiply by 0.5 because a VLI household makes no more than 50% of the state's median income.)

Maximum VLI family can pay toward housing = $50,500 x 0.3 = $15,150 annually

Difference between FMR and what VLI family can pay = $24,000 – $15, 150 = $8,850 annually

Subsidy given by HUD through local housing authority = $8,850 / 12 = $738 per month

Many local government assistance programs are based on similar criteria; however, the largest welfare programs are usually federally funded and based on the nonregionally specific federal poverty thresholds.

The U.S. Census reports several other measures of income inequality in addition to the poverty rate. One of these is the GINI Index, with a value of 1 representing prefect inequality and 0 representing perfect equality. There was no statistically significant change in the GINI between 2019 (0.465) and 2020 (0.469) (11). The Census also reports the share of household income by different income quintiles. This measure arranges household incomes from the largest to the smallest and then divides the incomes into five equal groups or quintiles. In 2020, there were 331 million people in the United States living in 129 million households for an average household size of 2.6 (331 million divided by 129 million) (13). Arranging the incomes for each of these households from the highest to the lowest and then dividing the incomes into five groups of 26 million households (129 million divided by 5 = 26) creates five income quintiles. Summing the total income in each quintile (or adding the household incomes for each 26 million) and dividing by the sum of all 129 million households gives us the approximate percentage of total household income for each quintile. Not surprisingly, in 2020, the lowest quintile had the smallest share of household income (3.4%), and the top quintile has the largest share (50.8%). In other words, the bottom 20% of households in 2020 accounted for just 3.4% of total household income, whereas the top 20% accounted for just over half (50.4%), though both quintiles contained an equal number of households (14). In 2020, the lowest quintile had household incomes of $27, 026 or less, whereas the households in the top quintile had incomes of $141,110 or more (14).

These disparities have only increased in the last 50 years. In 1968, when the Census started reporting this data, the bottom quintile accounted for 4% of total household income and top quintile accounted for 42.6%. In just over 50 years, the top quintile has increased its total share of the total household income pie by nearly 10 percentage points, and the share of the other four quintiles has shrunk (14). This is just one measure of the increasing income inequality in the United States. Similar trends can be found throughout the world, particularly in higher income countries where income disparities have increased by even more than in middle- or lower income countries (7). The OECD, for example, reports that the middle class in Germany, though comparable in size to other developed economies, has shrunk significantly since the 1990s (15). We discussed classification of countries by per-capita income in Chapter 4. Lower income countries are those with per-capita incomes (GDP divided by population) below $1,046; the middle income range is from $1,046 to $12,695; higher income countries are those with per capital incomes exceeding $12,695 (16). Least developed countries are a special case of lower income as they also have weak human assets (i.e., low infant and maternal mortality, low literacy, and low enrollment of girls in secondary schools) and are economically and environmentally vulnerable (i.e., remote, agricultural dependent, and vulnerable to natural disasters). The LDC status allows countries access to a range of trade, development, and other benefits through various international organizations such as the UN, WTO, and EU (22).

Extreme Poverty

Poverty thresholds, or the poverty line, vary by country and are based on income distributions and cost of living in the country. Usually, poverty lines are based on the minimum income needed to procure basic needs such as food, shelter, and clothing. However, variations in income and cost of living make it difficult to apply standards from a rich country to those below it. A middle-class income in a lower middle income country may be below the poverty line in an upper income country. The international poverty line was created in 1990 to set a standard below which any household would be considered poor (i.e., a poverty line applicable to all countries in the world). Researchers examined the poverty lines in the poorest countries, as these would also apply to those with higher incomes. To account for variations in cost of living, the researchers made purchasing power parity (PPP) adjustments to the poverty lines. Subsequent to the adjustments, the poverty lines in six of the countries were around $1 per day (24). This was the first official international benchmark for extreme poverty. The international poverty line provides a uniform measure of extreme poverty. It allows researchers, policymakers, and others to gauge changes in the number of extreme poverty worldwide over time. The historical data suggest that considerable progress has been made in reducing extreme poverty during the last 3 decades. The line was revised to $1.25 per day in 2005, $1.90 per day in 2011, and $2.15 per day in 2022 (24, 43). New daily thresholds for extreme poverty were also created for lower middle–income ($3.65), upper middle–income ($6.85), and high-income ($24.35) countries.

Based on the new thresholds, nearly 9.1% of people lived in extreme poverty in 2017. The regions with the highest rates were sub-Saharan Africa (38%) and South Asia (9.8%) (43). Today, there are between 700–800 million people living in extreme poverty. The COVID-19 pandemic likely pushed another 70–115 million back into extreme poverty. The pandemic undid much of the progress of the preceding decade. The global poverty rate had been falling before the pandemic and stood at 8.6% in 2018. The 1st year of the pandemic reversed these gains and increased the global poverty rate to 9.2%. This was the largest increase in poverty since the 1990s (17). As discussed in previous chaptesr, supply chain issues, inflation, rising interest rates, and the war in Ukraine exacerbated the impact of the pandemic. Given recent trends, it is unlikely that the goal of reducing extreme poverty to 3% of the world's population will be reached by 2030 (27).

Approximately 80% of those in extreme poverty are young, reside in rural areas, and do not have much education. About two thirds of the extremely poor are below 24, though they represent only 40% of the world's population (27) Those below 15 are the most disproportionately impacted by extreme poverty. Half of the world's poor are below 15, though as a cohort they are only a quarter of the world's population (27). Women and girls are also disproportionately represented among the poor, as are the uneducated. Seventy percent of those over 15 living in extreme poverty did not have an education or very little education. These trends are not surprising given that wage earners tend to be older, working-age males. Childrearing and household duties disproportionately fall to women, particularly in developing economies. The opportunity cost for girls and women is high as they often have to forgo education and wage labor to fulfill household- and childrearing-related duties.

Causes of Poverty

Poverty is seen as the result of different factors, some structural and others agentic (18). **Structure and agency** can be seen as antipodes, or opposite ways of conceptualizing poverty. Structural explanations typically focus on exogenous or external factors such as job availability, cost of living, educational opportunities, or systemic racism; agentic explanations place the onus on people and see poverty as the result of poor choices made by individuals. In lower income countries, particularly those that are least developed (LDCs), poverty is usually the result of structural factors outside an individual's control. However, there is much debate on the causes of poverty in developed economies with higher per-capita incomes and stronger labor markets. Poverty is sometimes seen as inevitable; some invoke a culture of poverty, and a few see it as part of a divine plan.

Like other social phenomena, poverty is socially constructed. Values and meanings are attached to poverty and poor people through social construction processes (20). Depending on the nation or region, these values and meaning can be different. Social constructions are important because they help frame issues and shape attitudes, including stigma, and public and policy responses to the issue, including whether low-income people "deserve" help and what type of help. The U.S. public typically ascribes more importance to internal or agentic factors in explaining poverty (21).

An important debate in the United States today is the relative importance of neighborhood and family in economic mobility. Some researchers suggest that neighborhoods are more important in predicting economic mobility, and others argue that while neighborhoods are important, family structure, parenting, and school quality may be more important. The debate is important because economic mobility in the US has plummeted from 90% to 50% in recent times and policy responses are driven by how poverty and its causes are framed (16). The land of opportunity, it seems, is increasingly less so, particularly for children born in certain neighborhoods or to certain groups.

Approaches to Economic Development

Poverty can be addressed through either increasing incomes or reducing the impact of poverty through subsidies for food, housing, health care, and other necessities. Income can be increased through direct cash subsidies, such as universal basic income (UBI), or economic growth, which impacts employment and wages. Though global capitalism, which includes world trade, has increased economic development across the world and reduced absolute poverty, income and wealth inequalities are growing, even in the most developed economies. Additionally, extreme poverty continues to persist, and many countries seem to be trapped in the middle-income range. Former communist countries that have adopted global capitalism after the fall of the USSR have had mixed success. Many of Europe's most languishing economies are former states of the Soviet Union. Russia, the torchbearer of the Soviet system, adopted a peculiar style of capitalism, sometimes called a *kleptocracy*, in which important state industries were transferred to a few oligarchs. The Russian economy is excessively reliant on the export of commodities, specifically oil and natural gas. The "strongman" legacy is alive and well as Vladimir Putin has maintained his grip on power for over 2 decades. The support given by the West to Ukraine recently has led some Russians to claim that the Ukraine conflict is the first salvo in an existential war against the West. As these examples attest to, the global capitalist system, despite its many successes, seems to have misfired in some important respects.

As WWII came to a close, Allied leaders met in Bretton Woods, New Hampshire, to create new international institutions to ensure trade, open borders, and macroeconomic stability. To this end, as discussed in Chapter 5, they created the World Bank, International Monetary Fund (IMF), and the World Trade Organization (WTO). These institutions reflected Keynesian thinking about the importance of government in managing economies. Chapters 4 provided an overview of both Keynesian, or demand-side, and monetary approaches to economic growth. The demand-side theories argue that government can stimulate aggregate demand, which is especially important during recessions. The monetarists, in contrast, viewed recessions as a natural part of the economic cycle and argued against any government intervention. Inflation was the monetarist's *bete noir*, the thing to be avoided. Their remedy was tight control by the government over the money supply. Fiscal policies use the tax system (tax deductions or credits) to incentivize certain types of expenditures; monetary policies, usually set by the Central Bank or Federal Reserve, control the money supply. Both have been used by governments to combat economic downturns and inflation. For example, the New Deal, crafted by the Roosevelt administration in response to the Great Depression, impacted every part of American society. New policies, programs, and institutions were created to provide employment, stabilize the financial industry, develop infrastructure, increase homeownership, increase government regulation and oversight, and provide a social safety net in the form of the Social Security programs. Similarly, during the 2007–2008 recession, the U.S. government used both fiscal (American Reinvestment Act) and monetary (quantitative easing) policies to stabilize the economy. More recently, in 2022 and 2023, the Federal Reserve increased its federal funds rate several times to reduce inflation. As the federal funds rate rises, banks increase their consumer lending rates, making borrowing more expensive. As debt becomes more expensive, consumers curtail expenditures, thereby reducing inflation, at least in theory; in the United States, inflation started easing toward mid-2023.

Keynesian ideas about the role of government spending in stimulating, and, if needed, borrowing to spend, eventually won over governments throughout the world. The "postwar Keynesian consensus" was supported by government, businesses, and labor. Many believed that poverty would soon be eliminated and prosperity would be shared by all (30). As countries threw off the yoke of colonialism, many adopted Keynesian ideas. However, its shortcomings were soon exposed. It became apparent that trade between rich and poor countries had the potential to impoverish poor countries if poor countries exported primary goods or commodities (e.g., rice, coffee) and imported finished goods from rich countries (this is the main argument made by dependency theorists). This was not much different than mercantile colonialism used by the British, most notably in India, and other colonial countries. To keep increasing imports of finished goods, poor countries would have to export larger amounts of commodities to avoid severe **trade imbalances**, the impact of which was discussed in Chapter 5. These "declining terms of trade" would allow richer countries to keep growing without commensurate growth in poorer countries; additionally, poor countries would need to exploit more and more resources for export, reducing the resources available for indigenous development (30).

To avoid this trap, developing countries needed to reorient their economies toward industrial production. **Modernization and structural** theorists offered different perspectives on the remedies that were needed to shift developing economies to production. Some saw investment, or capital, as the main barrier, others called for active state intervention to manage the transition to industrialization, and some saw traditional societies lacking the values and culture to successfully import capitalism. The modernists saw poorer countries as largely traditional, with value systems that were not geared toward technological or material progress. Tradition provided the template for living, with each generation largely replicating the ontologies, or ways of being, of their ancestors. These societies did not emphasize self-interest, accumulation, specialization, or technological growth like modern capitalist societies. Though traditional societies were spiritually rich, they never outgrew a subsistence lifestyle (31). The modernists argued that Europeans, like other societies, were mostly premodern for much of human history. Though scientific and technological advances occurred throughout the world, around 500 years ago innovation in Western Europe, combined with entrepreneurialism, eventually gave rise to the industrial age (31). The main impetus for this transformation, according to the modernists, was capitalism. As discussed in previous sections, specialization, competition, and innovation are all features of capitalist economies. Capitalism increases the general standard of living as wages, goods, and services all increase over time, leading to the "age of mass" consumption (31). Capitalism also engenders political and social change by upending older, traditional systems of power, as these can create significant market distortions. Representative government and individual freedoms grow, as these are necessary for capitalism to thrive. Modernists also argued that poorer countries could avoid many of the missteps made by Europeans and that richer countries should help the poorer ones through economic aid, technology, and access to their markets (31).

The **dependency theorists** did not share the optimism of the structuralists and modernists. To them, poor countries would not be economically uplifted through structural or cultural transformations as richer countries needed to keep them poor. Furthermore, the elite classes in poor countries would ally themselves with capitalists from richer countries to maintain the poverty of the masses. The rich countries needed a constant supply of cheap raw materials, including agricultural products. The elites in the poor countries would ensure its supply by using, or coercing, local labor to produce those materials needed for export in exchange for the wealth and privileges given by capitalists from richer countries. This created imbalanced growth as local economies were configured to provide materials for export with few resources allocated for local development. To the dependency theorists, the relationship between rich and poor countries was inherently exploitative and created poverty, which allowed for continued immiseration (31).

Other approaches to development include **autarky,** a state that is largely self sufficient and engages in little trade. No nation has successfully developed using autarky, though it could work to some extent on a smaller scale (e.g., communes). Developing economies in the 20th century turned to some combination of statism or using the state to marshal or leverage capital and resources for development and foreign investment and expertise to develop local industries. **Import substitution industrialization (ISI)** was a favored strategy by many developing countries. To foster the development of local industries, governments would impose quotas or tariffs (taxes) on imported goods or require a certain percentage of certain products to be locally produced (30). Governments would also license the manufacturing of certain products to a few or one company, thereby creating an oligopoly or monopoly. By adopting policies that favored local goods, governments sought to give local manufacturers time to accumulate capital and expertise to compete with the larger foreign manufacturers who were already producing higher quality goods for cheaper prices given their economies of scale. Many emerging economies adopted ISI as an economic development strategy. However, its impacts were uneven at best. Many foreign firms were able to skirt import tariffs by creating or buying local companies. Profits made by local producers were often channeled offshore to the parent companies. Governments sought to control capital flight through taxation and restriction, but this made other foreign investors wary of investing in the country. By restricting competition, ISI policies did not provide any incentive for innovation, efficiency, or product quality (30). Local firms often produced subpar goods that, while consumed locally, were not of sufficient quality to meet international standards. Additionally, the gains from ISI only benefited a small number of local producers and their employees, so no mass market was created for the consumption of goods and services (30). The agriculture sector was neglected as governments focused their attention on industrial growth. These imbalances led to a rural exodus as agriculture became increasingly less profitable. Lastly, ISI policies encouraged venal behavior as bribes were exchanged for manufacturing licenses, funds were skimmed from development budgets, and cushy government jobs were handed out to friends and family instead to those most qualified (30).

ISI policies required heavy government intervention in markets. Not surprisingly, the rejoinder came from those who favored no government interventions. The neoclassical theorists rejected development economics as a separate discipline and urged a return to classical economic ideas about markets and trade. They argued for less regulation, taxation, and protectionism. In other words, countries would develop by embracing capitalism wholeheartedly and opening their borders for trade. Initially, this would be painful to local producers as they would not be able to match the prices and quality of larger foreign producers, but, over time, local economies would develop and flourish. Some government intervention in the form of subsidies was acceptable as long as it supported local producers making export quality goods. Contrary to the arguments made by dependency theorists, neoclassical proponents argued that trade with rich countries would contribute to knowledge and skills transfer and enrich poor countries in the long run.

This view led to what were loosely called **structural adjustment policies (SAPs)** that aimed to undo the protections, tariffs, licenses, and other market distorting programs and policies implemented by developing countries. Unfettered markets were at the center of structural adjustment programs. To this end, industries were privatized, trade restrictions were reduced, foreign investment was encouraged, currencies were devalued, and many parts of the economy were deregulated. Structural adjustment also included **fiscal austerity** measures to slash government expenditures, which also reduced spending on many social service or antipoverty programs.

The neoclassical view slowly gained traction and became a precondition for aid by many developed nations and **international financial institutions (IFI)** such as the IMF and World Bank. In 1989, economist John Williamson coined the phrase "Washington consensus" to describe the structural adjustment policies espoused by those seeking to address the economic problems in Latin America. The phrase stuck and was used to describe SAPs and other austerity measures imposed by IFIs on countries that sought their help. Many of these reforms did not achieve economic growth, and some misfired spectacularly, particularly in sub-Saharan Africa. Similarly, countries in Eastern Europe or the former Soviet Union states that implemented at least some portion of the Washington consensus grew very little (32). In general, many developing countries following the Washington consensus model fared badly during the 80s and 90s. Some called this the "lost decades" for developing countries (33).

In contrast to the poor growth of those following structural adjustment prescriptions, some countries, like China and Vietnam, which transitioned to a market-based economy on their own terms, performed well. The growth in the Chinese economy, in particular, has been nothing short of spectacular. In the 4 decades between 1980 and 2020, China lifted nearly 800 million people out of **absolute poverty** by focusing primarily on economic development (28). The factors causing China's rapid growth are still not fully understood. Market-based reforms, privatization of former state industries, low fertility, a large resource base, the availability of a a large labor class, and competent leadership were all important factors. China opened its economy to trade and introduced market-based reforms starting in the late 1970s. China first reformed its agricultural sector and then fostered the development of low-skilled, labor-intensive industries. As agriculture became more efficient, less workers were needed. Those leaving agriculture were able to find work in the burgeoning industrial sector. This led to greater urbanization and **remittances**, or money sent back to the villages, further boosting rural incomes (28).

Initial productivity improvements in capital and labor, also called **total factor productivity (TFP)**, played an important role in China's rapidly growing GDP. Productivity improved when workers moved from state-owned to private industries in the 1990s, when there was little incentive for increased productivity (29). These improvements have since slowed down, so it is unlikely they will contribute to economic growth in the same manner in the coming years. Joint ventures with foreign firms were also pivotal in speeding up the transfer of knowledge and technology, which initially increased productivity significantly (29). As some of these transfers were the result of intellectual property theft or piracy, the productivity impacts of joint ventures with foreign firms are also likely to attenuate, or lessen, over time. Other large gains in the economy were likely due to the addition of more capital and workers, or **factor accumulation** (29). The addition of low-cost labor, many from rural areas, as well as skilled workers (labor), through expansion of education, along with significant capital (money) investments by the government and private sector, including foreign direct investment, was likely the biggest factor in GDP growth in the past 40 years. As the Chinese economy matures, it can no longer rely on factor accumulation or productivity growth from the same sources (privatization, knowledge transfer from foreign partners) to the same extent. Instead, it must invest in human capital to increase the proportion of skilled workers, especially as rising incomes have made labor more expensive. China must also encourage more domestic consumption and become less agriculturally focused, more urbanized and suburbanized, and less reliant on an export driven economy (29). The role of the state must evolve as heavy state intervention creates distortions, as discussed in previous chapters. The factors powering China in the coming years will likely be different than those that impelled its rapid economic ascendancy in the recent past (29).

The mixed success of the major economic development ideas of the past century shows that there is no one road to economic prosperity. The modernization path of the United States, though similar to Europe, was also quite different in some important respects. As discussed in previous sections, World War II played a decisive role in cementing American economic dominance in the 20th century. Many Latin American countries were well on their way to middle income early in the 20th century, but many also became mired in the "middle income" trap. Countries in sub-Saharan Africa have had the least success, though there have been some notable exceptions like Botswana and Namibia. Wars, low population density, and resource mismanagement have all played an important role in under-development in the region. In East Asia, outside of China, the biggest success stories have been South Korea, Hong Kong, Taiwan, and Singapore, which today are advanced developed economies. Their ingredients of their success are still debated; however, their strong focus on an export-oriented development was a major factor, along with significant investments in infrastructure and education. Foreign aid also likely played an important role. South Korea may have also benefitted from a low dependency ratio (i.e., a demographic dividend) as it had many working-age people but not as many children or elderly to care for. After independence, India, the largest of the South Asian countries, chose import substitution (ISI) polices and heavy state involvement in the economy, along with central planning that was characteristic of the former USSR. India, however, also liberalized during the early 1990s as per the SAP requirements of the IMF and World Bank to whom it had turned for loans. Since then, it has become one of the fastest growing economies in the world, though it is home to the largest number of people living below the extreme poverty line.

There are as many stories are there are countries in the world. Today, economic development prescriptions eschew, or reject, a one-size-fits-all approach. Protections for the poor and the environment are not seen as secondary to economic development but as important as economic development goals and necessary for long-term, sustainable development.

Poverty Policy in the United States

The United States is often seen as less progressive when compared to Western European or Scandinavian countries. U.S. antipoverty programs, with some exceptions, are usually limited to **dependents**, or those who are seen as unable to work because of age or disability. The U.S. social safety net is, however, extensive and includes cash assistance, subsidies, income tax-related benefits (credits or deduction), low interest loans, loan forgiveness, credit enhancements, direct provision of goods, insurance, stipends, grants, and legal services. Eligibility criteria include **means testing** or evaluating eligibility based on some combination of income and assets. For most programs, the federal poverty lines (FPL) or some multiple of them (e.g., 150% of FPL) are the basis for eligibility. Most federal programs are limited by the annual amount appropriated for them, so qualification does not mean automatic provision of **government aid**. Some programs, such as Supplemental Security Income (SSI), Section 8 (housing vouchers), and the earned-income tax credit (EITC), are not based on the FPL but use other methods for means testing. The assumptions used to create the federal poverty line, or income level, for various household sizes were discussed earlier in this chapter. Since 2011, the Census Bureau has published data for the supplemental poverty measure (SPM), which includes benefits provided by government antipoverty programs into consideration, as well as family composition, regional differences in housing costs, and expenses for food, clothing, and utilities. The supplemental measure provides a more realistic assessment of need as it is regionally sensitive, accounts for basic expenses other than food, and includes assistance received from the government when measuring a family's income or resources (36). In 2018, the SPM poverty rate was about one percentage point higher than the official poverty rate; the largest increase in poverty (1.4 percentage point increase) occurs for those without a high school diploma when poverty is assessed using SPM criteria (36). This underscores the impact of education in upward socioeconomic mobility.

The two biggest antipoverty programs in the United States are Social Security and the EITC. Together, these programs kept over 36 million people out of poverty in 2018 (36). The EITC is the biggest tax-based program for low- and moderate- income families and provides a tax credit to eligible households that phase out at higher incomes. The tax reductions provided by the EITC helped keep nearly 8 million out of poverty in 2018 (36). Social Security, the biggest cash assistance program in the country, kept over 27 million, the majority of whom were elderly, out of poverty in 2018 (36). Participants pay into the Social Security program during their working years through payroll taxes and become eligible for regular monthly payments at retirement based on their contributions. The program also has survivor benefits for spouses. The solvency of the Social Security program, as discussed in Chapter 6, is under threat due to declining fertility. The SSI program, funded through general tax revenues, is another means-tested cash assistance program for the income and asset-limited elderly, disabled, or blind. Unemployment benefits, which are usually administered by the states and paid for through employer taxes, include weekly cash assistance for up to 26 weeks and training- and education-related benefits. Eligible families received food assistance through the Supplemental Nutrition Assistance Program (SNAP), which provides debit cards for food and related items. The federal government provides funds for cash and services to needy families to states as a block grant through the Temporary Assistance for Needy Families (TANF) program, created in 1996 as part of the Clinton administration's Personal Responsibility and Work Opportunity Reconciliation Act (PROWRA), which reformed welfare to reduce dependency and encourage work and two-parent households. In 2020, $31.6 billion was allocated through TANF. Of this amount, only $7.1 billion was given as cash assistance (35). The TANF program, while large, has lost 40% of its original value as it has not been adjusted for inflation or population increase since it was created (35).

The U.S. medical system is fragmented, leading to significantly higher health care expenses than other developed nations. The United States spent $3.6 trillion, or almost $12,000 per capita, on health care in 2018, more than any other nation (39). The Patient Protection and Affordable Care Act (ACA), known colloquially as Obama Care, expanded insurance coverage significantly. The U.S. Department of Health and Human Services estimates that over 30 million people received medical coverage because of the act's provisions. Most of those with new medical coverage enrolled in Medicaid, one of two federally funded health insurance programs for the general population (37). Medicare, the other large program, is tailored to those over 65. MediCaid/CHIP (Child Health Insurance Program) and Medicare together provided health coverage for over 165 million Americans in 2021 (38). The federal government also provides health care access for military veterans through the Veterans Administration (VA).

Many other benefits are provided through all levels of government (federal, state, county, and local). U.S. antipoverty policy continues to change depending on the zeitgeist and administration in power. They have made considerable progress in some arenas, however, the economic growth in recent decades has not been shared by all. The basic premise of a meritocracy is an equal playing field, but many would argue that one has never existed in the United States and that the social safety net, while extensive, is a long way from creating one.

One approach to poverty that has recently generated a lot of buzz is the provision of a UBI, or cash, to all adult citizens. Unlike other forms of support, UBI would be provided without any strings attached. This would also reduce the bureaucratic burdens associated with determining eligibility or tracking compliance. Some areas, including a few in the United States, have tried or are currently experimenting with UBI. Finland implemented a randomized experiment replete with a control group to study UBI. The findings seem to indicate modest positive effects as there were small increases in employment and larger increases in measures of well-being compared with the control group (47). Additionally, those receiving UBI were more likely to seek unpaid work or training, which further boosted confidence and feelings of well-being. Those getting UBI also reported greater trust in people, government, and other societal institutions (47). The City of Stockton in California also experimented with UBI. Preliminary findings suggest that the $500 cash allowance increased full-time employment among recipients, who also reported an easing of psychological distress (anxiety, depression) associated with economic uncertainty. UBI's detractors argue that long-term income guarantees will create huge disincentives to work; create entrenched, generational poverty; and reduce individual initiative, labor productivity, and economic growth. Furthermore, some people may not need the income and the government would essentially be subsidizing them by taking needed resources from the poor. The policy would also be difficult to finance given its substantial, ongoing costs.

Conclusion

Global capitalism has undoubtedly increased the material standard of living as the goods and services provided through industrialization, specialization, and trade have made life easier. Technological growth has also been substantial, increasing both our material well-being and our understanding of our world and universe. However, there is a significant cost. The links between capitalism, resource consumption, and waste have been previously discussed. In addition, many of the health problems, including mental health, can be traced to the lifestyle changes engendered by capitalism (e.g., sedentary lifestyles, processed foods, auto dependency). Its proponents, however, argue that no other economic system has similar productive potential given that humans are primarily self-interested and act accordingly. Other systems, particularly those that seek to address inequality, are often dismissed as unrealistic or impractical. The central idea of liberalism is that the social good arises only when individuals are allowed to compete in free and fair markets. It cannot be enforced or mandated but must be the result of millions of actions in the various markets of society. The experiences of centrally planned economies lend credence to these assertions. However, poverty and relative inequalities (e.g., health, education, or mobility) continue to persist despite considerable effort to ameliorate both. Some argue that a certain level of inequality is inevitable and serves as a motivation for those who wish to climb the SES ladder. Others find extreme inequality and its impacts untenable, especially in a world that produces enough necessities for all. Additionally, the waste in developed countries and by the wealthy in poor countries attest to the excesses and inequalities generated under global capitalism. The climate change impacts, as discussed in earlier chapters, are also undeniable and likely to worsen. As poor countries move up in income, many, if not all, will pollute and waste in ways similar to developed countries today. Additionally, the impacts of AI are uncertain and potentially disruptive, causing widespread un- and underemployment. It is not implausible that in the not-too-distant future large segments of society will not be productively engaged in any type of work, as there will not be enough work, and will receive UBI and other forms of assistance, some of it provided by AI (e.g., AI providing primary care or 3D printed basic housing).

Some questions to consider, as we look to the future, include these: Can the world achieve similar levels of technological progress in a noncapitalist system? If poverty is created by capitalism, what are alternative ways to organize the economy and society (as the inequality and poverty we see today did not exist in societies without the same type of property ownership that exists under modern capitalism)? What are some of the other ways to define progress that account for social, psychological, and spiritual growth? Can capitalism be reconciled with the goals of sustainability?

Chapter 8

Public Health

Tracking and Measuring Population Health and the Social Determinants of Health and Disparities in Health

Key Terms

Mortality: Statistics related to death. All-cause mortality rate is estimated by dividing the dead in a region in a particular time period by the region's total population. Mortality rates help us understand the severity of a condition and evaluate the effectiveness of interventions. We can also use differences in mortality rates between groups to identify health disparities, an important measure of equity.

Morbidity: The impact of illness measured in terms of intensity and duration of symptoms. Morbidity can be measured for diseases, injuries, and disabilities. Related concepts include disability-adjusted life years (DALYS), or years in perfect health lost, and quality-adjusted life years (QALYS), or years in perfect health gained. Both measures help us understand how diseases impact quality of life so that we can make decisions about the cost-effectiveness of different health interventions. Like mortality, group differences in morbidity can help identify health disparities.

Incidence: The number of new cases of a condition in a region in a particular time period. The incidence rate for a disease is the number of people who contract the disease in a given time period divided by the total population at risk.

Prevalence: The number of people who have a condition in a region at a certain point in time. The prevalence rate for a disease is the number of people with the disease (i.e., the cured or dead are excluded) at a certain point divided by the total population at risk.

Surveillance: The monitoring of health indicators to identify new threats and their progression and outcomes.

Epidemiology: Epidemiologists are interested in describing the characteristics of a disease with respect to its origins, incidence, spread, prevalence, treatment and containment, and group differences in how the disease is experienced.

Social determinants of health: Social factors such as SES, education, health care access, and neighborhood- or community-related factors (e.g., crime, pollution) that impact disease exposure, response, and outcomes.

Health disparities: The disproportionate disease risk and harm experienced by certain groups due to social, biological, and other factors.

General plan: A plan guiding the development and land use decisions in a city within the context of certain elements like housing, circulation (transportation), safety, noise, conservation and environmental justice.

Other key terms will be in **bold** and defined in the text.

Review Questions

- What is a health disparity, and why is it important? How is environmental justice linked to health disparities?
- How do we measure the presence of a disease and its progression (including sickness) and outcomes (including death)? How can these measures help us identify health disparities?
- What are the functions of public health?
- What are some examples of the social determinants of health? How do social determinants help us identify and address health disparities?
- What are the different ways land use and public health intersect?
- How can urban planning help address public health related issues?

Introduction

Health and sickness can be difficult to identify. Health is sometimes seen as the absence of sickness, an unwanted condition of the body. In this sense, sickness is subjective. The passionate disagreements about COVID, its impacts, treatments, and vaccines illustrate the importance of interpretation and perception. These are not trivial considerations as they impact the social construction of illness and health. Consequently, certain constructions of illness and health are seen as legitimate and others as fringe, quackery, or absurd. Foucault, in the *Birth of Clinic,* discussed the development of the medical gaze and the power vested in physicians and allied professions to label certain bodily states as disease and others as healthy. The modern medical professions claim objectivity based on positivist methodologies, or the scientific method, to differentiate themselves from other claimants to healer. This medicalization of conditions has given the medical professions considerable power and established the primacy of biomedical perspectives on disease, illness, and health among the general populace (1). The list of conditions to which the medical domain lay claim expands, as does the medical industry, which includes the pharmaceutical companies and a vast number of allied health and wellness professions. The same forces that drive other sectors of the economy—specialization, market share, and growth—also shape and propel the medical industry. Some argue that medicine's primary mission to heal and improve quality of life has become secondary to its political economy; medicine is now another cog in the machinery of global capitalism. The proponents of modern medicine and public health, however, would argue that advances in medicine and public health, especially vaccinations, coupled with standardization of care, have resulted in an increase in life expectancy and reductions in both morbidity and premature mortality. In other words, people are living longer and the duration, intensity of sickness (**morbidity**), and death from sickness (**mortality**) have all decreased. By subjecting the healing professions to the standards of positivism, in which theories are tested, data is collected, and hypotheses are evaluated, modern medicine and public health have built a corpus of knowledge to treat individuals and communities.

These differing perspectives on health and health care in the modern age all contain a kernel of truth: People are living longer, and illness is better understood and treated; health care, however, has become a veritable industry that pushes a biomedical view of health and increases reliance on experts and the products of the health care industry. Additionally, the importance of perception in dealing with illness is given short shrift despite the very real presence of a placebo effect, an improvement in health driven by anticipation of the efficacy of a treatment. Individuals in clinical trials sometimes show improvement without any real medical intervention as they believe they are receiving treatment. The placebo effect is real and powerful. Often medicines touted as effective treatments don't have a large effect size: They don't improve patient outcomes much more than a placebo.

The political economy of modern health care increases market dependence and resource consumption, often to the detriment of the environment and sometimes the patient. Additionally, equity issues are rife in health care as health care access can vary by income, race, and other characteristics. In poor countries, the malnourishment of those at the bottom reduces immune response, increasing susceptibility to diseases and time to recovery. The poor also have reduced access to the tools and therapies of modern health care. In the United States, race, ethnicity, and income independently predict poor health outcomes for certain population groups. The differences in outcome are the result of **social determinants of health** such as income, health care access, education, community context (e.g., exposure to violence), and built-environment factors (e.g., food availability, air, and water quality) (13). The impact of social factors can be profound, as these are the first steps in the causal chain to ill health and mortality. Modern medicine primarily treats the proximate causes of illness, though the root causes are often structural, such as greater exposure to environmental hazards or community violence, and would require lifestyle changes, such as less processed food consumption or greater mobility (i.e., the ability to relocate, the ability to physically move or exercise), which are often difficult or impossible to achieve because of individual and social constraints.

Health and related industries are the biggest area of expense in the United States and most advanced economies. In 2019, $8.5 trillion was spent on health worldwide, or about a tenth of the global GDP (3). Developed countries were the biggest spenders, spending an average of $4,500 per capita (3). Among developed countries, U.S. spending was the highest. The United States spent over $4 trillion on health care in 2020, or $12,530 per capita (2). Nearly a fifth of the nation's GDP was spent on health care. In 2019, before COVID, health care accounted for 17.6% of the GDP. In 2020, nearly a third (31%) of health care expenses were for hospital care, with another fifth (20%) going to physician and clinical services (2). Spending on prescription drugs accounted for 8% of total spending. Americans also spent $30 billion on alternative health care services such as chiropractic care or acupuncture and products such as herbal supplements (4). These expenses are likely to increase with the aging of the baby boomer cohort. Slowing fertility in the United States also means fewer working-age people will contribute to the public and private insurance schemes that defray most health care in the United States. The impact of low fertility on population age structure was discussed in Chapter 6.

It's important to evaluate the impact of increasing medical expenses on health. To do this, it's necessary to identify relevant metrics, or measures. Mortality, or death-related, and morbidity, or sickness-related, measures are often used to describe illness progression. Population morbidity and mortality data are collected for common and serious illnesses in developed countries. The lack of morbidity and mortality data for various health conditions in poorer countries makes it difficult to assess needs, evaluate the impact of public health programs, and identify disparities in outcomes. The collection and analysis of health data is a core function of public health.

For many years, policymakers believed that improvements in the economy would eventually lead to better health outcomes. This view led to a disproportionate emphasis on economic development. While this is true, causality can also run the other way: Improvements in health can increase economic progress. Excessive morbidity, or sickness, can reduce the number of days at work and productivity. Healthier workers take fewer sick days and are more productive when they are at work as they have more energy and mental acuity. The UN's 17 SDGs include "healthy lives and well-being for all ages." In addition, ending world hunger is a separate goal (5). Though the world produces enough food for all, the UN estimates one in 10 people are suffering from hunger (5). Hunger and malnutrition are a significant cause of infant and child mortality: nearly half (45%) of the deaths among children under 5 are linked to malnutrition. The absence of hunger is a precondition to good health; however, it does not guarantee it. It's possible to consume the minimum daily calories without appropriate macro (proteins, fat, carbohydrates) and micro (vitamins, minerals) nutrients. In the United States, for example, low-income individuals are more likely to be overweight and obese as processed foods high in sugars, fats, and other additives are more readily available in low-income neighborhoods. It's cheaper to buy chips, candy, and hamburgers in many poor neighborhoods than a salad or a fruit bowl. These neighborhoods are **food deserts**. The modern world, paradoxically, is afflicted with the comorbidities associated with both excess weight and underweight: Nearly 2 billion adults are overweight or obese, and 460 million are underweight (6). Many developing countries, given their stark income disparities, must devote significant resources to combatting both issues.

This chapter provides an overview of basic public health functions, such as disease surveillance and epidemiology, and concepts such as incidence, prevalence, morbidity, and mortality. Diseases have proximate, or immediate causes, which are medically treatable and more difficult to address, and underlying structural causes, such as poverty, greater risk of exposure to environmental hazards, or lack of health care access. The chapter explores the link between various social factors and health to show that health care costs can be reduced and health outcomes can be improved with a stronger focus on the underlying causes of sickness. The consequences of the current health care regime on health disparities in the United States are discussed, as are the disincentives created by "medicalization" to address underlying causes of ill health. In this context, the chapter examines how the political economy of modern medicine, which includes "big pharma," relies heavily on treatment and management rather than prevention. The chapter also provides an overview of the urban planning and public health nexus by looking at how land use policies often exacerbate health outcomes and the planning tools available to improve population health. The chapter ends by examining how globalization shaped various aspects of the COVID-19 pandemic and the disparities in infection risk, morbidity, and mortality.

Public Health Concepts

Clinical medicine is concerned with individual health outcomes. In contrast, public health's **unit of analysis**, or the level at which the analysis is conducted, is the group. **Incidence** provides an estimate of the number of new disease cases in an area in a certain period. Incidence can be expressed as a proportion or rate (19). An **incidence propor-tion** is the number of new cases in a certain time period divided by the total population at the beginning of the time period (19):

Incidence proportion condition A = number with condition A in 2022 = 10 = 10
Total population at beginning of 2022: 100

The numerator for **incidence rate** is the same as proportion, but the denominator is the total period for which each person in the group was observed (19). For the same example, if 100 people were followed for 2 years and 15 were diagnosed with condition A in that period, the incidence rate would be

Incidence rate condition A = number with condition A in 2022 = 15 = 7.5%
Total time population was observed: 2yr x 100

Prevalence estimates the relative presence of a condition in a population group of interest by dividing the total number of new and existing cases by the total population (19). For example, if 20 people had condition A in 2022 and 10 new cases were reported in 2022, the prevalence for a population of 100 would be estimated by the addition of the new and old cases (10 + 20 = 30) and dividing by the total population (30 /100) for a prevalence rate of 30%. Incidence and prevalence, like other public health metrics, can be measured for the whole population or specific population cohorts or subgroups (e.g., age groups, at-risk groups, etc.). Similarly, data can be reported at any geographic level, such as the neighborhood block, census tract, city, state, nation, or the world.

Public health includes many different specialists with the common purpose of maintaining population health. Practitioners include clinicians such as physicians and nurses; program planners who plan and manage complex population health programs; health educators and promoters who design and implement programs to increase population **KAPs** (knowledge, attitudes, and practices); hospital administrators; epidemiologists who study disease origins and spread; civil engineers and others concerned with environmental contaminants; and data experts such as biostatisticians. The WHO lists 12 essential public health functions, and the U.S. Centers for Disease Control and Prevention (CDC) subsumes these into 10 essential services (7, 8). These can be grouped under health surveillance, epidemiology, health system management, health promotion, and disease prevention. In the United States, emergency management, which includes responding to terrorism, has been included in public health's purview.

Public health **surveillance** depends on the entire health care network to monitor health indicators and identify emerging health-related threats, their progression, and their outcomes. Effective surveillance requires coordination between many different public and private entities for data collection, analysis, and dissemination of findings. The CDC, which is part of the U.S. Department of Health and Human Services, is the primary public health organization at the national level in the United States. The CDC's MMWR (*Morbidity and Mortality Weekly Report*) is used by the agency to disseminate and discuss public health data and recommendations. (The MMWR is sometimes called "the voice of CDC.") Each state also maintains a public health agency, and there are 2,459 local health departments, of which the majority (1,886) are administered by county or city governments (11). Public health data is collected at the local level and aggregated at various levels of government to provide a snapshot of population health at local, state, and national levels.

Epidemiology is concerned with the origins (causes, risk factors), spread (distribution), and containment of conditions impacting health (9). These conditions can include infectious diseases (flu, food poisoning); illness due to environmental hazards (heavy metal poisoning, air pollutants); injuries (homicides or suicides); noninfectious conditions with a significant health impact (cancers); natural disasters; and terrorism (11). Occasionally, a film with public health overtones (e.g., *Outbreak*, *Contagion*) will become popular and bring public health into the national conversation; however, public health discourse tends to be limited to specialists. In contrast, there have been many shows about clinical medicine. The measures taken to stem the spread of COVID-19, including public shutdowns, masks, and vaccinations, generated heated public debates, as did the official morbidity and mortality statistics. Significant numbers of people, particularly in the United States, objected vigorously to the official recommendations and questioned data on incidence, prevalence, morbidity, and mortality. Many argued that risk was overblown and limited primarily to the elderly and those with underlying conditions. For some, the efficacy of public health measures such as masks was questionable; others viewed public health requirements such as vaccinations as a direct attack on their freedoms.

Public health's relative obscurity prior to the COVID-19 **pandemic**, a country or worldwide disease outbreak, is perhaps best explained by its focus on **primary prevention**. Through vaccinations, health promotion and education to alter risky behaviors, and regulation, including banning of known harmful substances, primary prevention seeks to reduce disease incidence. This is also the most cost-effective of all health approaches, but the impacts are harder to prove as the condition has not occurred. In contrast, clinical medicine often focuses on **secondary and tertiary care,** which occurs after an illness or trauma has occurred. Secondary health care includes health screenings for early identification of disease onset and monitoring of progression. Examples include regular health check-ups (e.g., blood pressure or breast cancer screenings). Secondary care can also include management through medication, therapies, and lifestyle changes to prevent escalation. In some cases, secondary care can result in a return to a disease-free state (e.g., exercise and diet program to eliminate type 2 diabetes). Tertiary care involves the management of complex, chronic conditions to minimize symptoms and maximize quality of life (e.g., rehabilitation programs, drugs for managing chronic diseases such as arthritis or depression).

The standardization and quantification of assessment and outcome criteria occurred in lockstep with the changes in epistemology, or how we know things, which occurred during the Enlightenment in the 17th and 18th centuries. **Epistemology** sets the framework or rules for collecting data, analyzing data, and making inferences, the conclusions supported by the data (14). The scientific method in which theories are tested empirically through the collection of data, both qualitative and quantitative, is the bedrock of **Enlightenment** epistemology, which emphasizes reason and data that can be verified by the senses or instruments. Enlightenment philosophy upended tradition, the *ancient regime* comprised of the Church, monarchy, nobility, and epistemologies that privileged religious knowledge. The scientific method produced advances in science and technology, including medicine, which helped establish its dominance over prior epistemologies and provided the rationale for some of Enlightenment's most trenchant critiques of existing paradigms (e.g., the geo- or Earth-centric view of the universe). In medicine, this era saw the development of germ theory, which posited that microorganisms were responsible for disease. Previously, **miasma**, or noxious vapors emanating from rotting organisms, was implicated in the spread of disease.

Health and allied fields are concerned with reducing sickness and preventing premature death. Establishing death, however, is sometimes difficult as cessation of the circulatory system, which includes the heart, and neurological cessation, which includes the brain and brain stem, don't always occur at the same time. Neurological criteria are commonly used, specifically the irreversible death of the entire brain. However, in rare cases biological resiliency is noted despite the cessation of all brain activity (15). The advent of mechanical ventilators has confused matters more as adequate blood **perfusion** is maintained even when the patient's brain shows no sign of activity (i.e., the patient is brain dead). The appropriate time to remove organs from organ donors is another major issue as the window for preserving viable organs is small (15). First responders have the unenviable task of balancing lifesaving interventions with preserving viable organs as cardiopulmonary resuscitation (CPR) can maintain adequate oxygenation for sometimes as long as 30 minutes after the heart stops. Death or **mortality rates**, as discussed in Chapter 6, can be estimated for all causes or specific health conditions. Like other rates, they are calculated by dividing the number who have died from the total number who could have died (crude death rate) or number who have died from a particular health condition divided by the total number with the health condition.

As shown in Tables 8.1 and 8.2, the leading causes of death vary depending on level of economic development. Many low-income countries must still contend with premature deaths due to **communicable (or infectious)** diseases such as tuberculosis (TB) that are transmitted from another person or **vector-borne** conditions such as malaria, which are carried by another organism such as a mosquito. Poor transportation infrastructure, traffic control, and enforcement of traffic laws also result in a high number of preventable deaths on the road. Low-income countries also have much higher rates of infant and child mortality due to diarrheal and other infectious diseases linked to poor living conditions and environmental hazards.

Causes of Death	Number of Deaths
Heart disease	696,962
Cancer	359,831
COVID-19	350,831
Accidents	200,955
Stroke	160,264
Chronic lower respiratory diseases	152,657
Alzheimer's	134,242
Diabetes	102,188
Influenza and pneumonia	53,544
Kidney diseases	52,547

Adapted from "Mortality in the United States 2020" (NCHS Data Brief, 427), by S. Murphy et al. (59).

Table 8.2 Leading Causes of Death, High- and Low-Income Countries (2019)

Worldwide	High Income	Low Income
Heart disease	Heart disease	Neonatal conditions
Stroke	Alzheimer's	Lower respiratory infection
COPD	Stroke	Heart disease
Lower respiratory infection	Trachea, bronchus, lung cancers	Stroke
Neonatal conditions	COPD	Diarrheal disease
Trachea, bronchus, lung cancers	Lower respiratory infection	Malaria
Alzheimer's	Colon and rectum cancers	Road injury
Diarrheal disease	Kidney diseases	Tuberculosis
Diabetes	Hypertensive disorder	HIV/AIDS
Kidney diseases	Diabetes	Liver cirrhosis

Adapted from the WHO "The Top 10 Causes of Death," December 9, 2020, https://www.who.int/news-room/fact-sheets/detail/the-top-10-causes-of-death.

Morbidity is concerned with disease impact, which includes disease symptoms, or the immediate signs of disease presence, and **sequalae**, the ongoing impacts of an **acute**, or short-term, condition. Symptoms of COVID-19 that persist after the initial disease has subsided are examples of sequalae. Diseases, trauma, or drug use can also cause psychological symptoms and sequalae.

Morbidity, like mortality, is sometimes difficult to identify. Many illnesses have objective markers that can be measured independently of patient perception; however, others lean more on self-reported symptoms. An example is fibromyalgia, which results in chronic pain and fatigue (17). Psychologically or physically stressful events can trigger the condition, or sometimes there is no single precipitating event. The condition cannot be cured and is managed through medication and lifestyle changes that include exercise and stress reduction (17). In some cases, conditions have no discernible symptoms and can only be ascertained through a diagnostic test. Hypertension, or high blood pressure, is one such example. The condition increases heart attack and stroke risk and is evaluated regularly at physician visits and annual physical examinations (16). Nearly half of U.S. adults are believed to be hypertensive (16).

Additionally, disease etiologies are complex. Multiple factors, including SES, environmental exposures, food consumption practices, and health care access, contribute to the onset of disease and progression. Treating a disease solely in terms of its **biomedical**, or physiological or pathophysiological, characteristics can be reductive, or overly simplistic, and often doesn't lead to long-term health. Chronic conditions such as diabetes, heart disease, and cancer are the leading causes of morbidity and mortality in developed countries and among the well-off in poorer countries. Lifestyle factors, specifically food, exercise, and stress, are implicated in their onset. Physicians counsel their patients to eat healthier, exercise, and reduce stress as these can sometimes slow disease progression or reverse it; however, many simply rely on medical interventions such as medicines and surgery as these dominate the discourse on health and wellness.

Medicalization is the process by which parts of the social world are brought under medical or expert control (18). The process engenders new ways to profit from diseases that can only be treated properly through biomedicine. For example, stress, birthing, weight, and substance abuse have all been medicalized to some extent (18). Medicalization serves the political economy of capitalism, of which assumptions, values, and concepts help create the markets for medical goods and services. The social construction of health in the modern era often treats the disease and not the person. The emphasis on curing or managing primarily through the medical system leads to disproportionate expenditures on hospitals, clinics, ambulances, drugs, and considerably less on factors such as food, environment, or stress that increase disease risk (18). In the next section, we look at the nonclinical, social determinants of health, which include poverty, environment, community, health care access, and education. These factors impact each other and health through many different pathways. **Addressing these factors will reduce the overall health care burden on society but also negatively impact those profiting from the current system.**

Social Determinants of Health

The WHO constitution states, *"Health is a state of complete physical, mental, and social well-being and not merely the absence of disease or infirmity" (19).* The absence of disease is a precondition to good health but does not guarantee it. Mental well-being can be elusive even when the individual is clinically healthy. Anxiety and depression are the most common mental health issues (22). About one in eight people in the world live with some type of mental health condition (22). Unfortunately, in some cases, mental health issues can lead to suicidal ideation (thoughts and plans) and attempts. Globally, the suicide rate is nine per 100,000, with the highest rates occurring in Europe (13 per 100,000)

(23). The higher incidence of suicide in Europe is primarily driven by the higher rates in Eastern Europe among former Soviet Bloc countries (23). In the United States, males (22 per 100,000) are four times as likely to attempt suicide as females (5.5 per 100,000). Also, in the United States, the elderly (75+), and young adults (25–34) are most at risk for suicide (24), as are Native Americans and non-Hispanic Whites. Throughout the world, between 700,000–800,000 people take their lives every year, though as many as 20 times that number attempt suicide (21).

Health states are a continuum and not antipodes of good or bad health. Many with identifiable illnesses or infirmities can be physically able and healthy; similarly, many with diagnosed mental health conditions can have mental health well-being some or a lot of the time. Both domains, physical and mental, can improve or exacerbate conditions in the other. The effect of exercise on a range of physical health indicators is well established (74). Exercise also plays a role in preventing mental illness and is therapeutic for a range of mental health conditions (72, 73). Regular exercise, for example, helps reduce the risk for chronic diseases such as heart disease and diabetes by reducing weight and improving other markers of health, such as blood pressure, cholesterol, triglycerides, and sugar. These improvements can help elevate mood, increase self-esteem, and improve overall mental health. Exercise impacts mental health directly by improving blood supply to the brain, increasing neural plasticity, and reducing cravings associated with different addictions (26, 28). Given its wide-ranging impact, exercise is recommended along with medications to manage a variety of physical and mental conditions. Though clinicians incorporate recommendations for exercise, their efforts are often perfunctory as follow-up is mostly focused on the effects of the prescribed pharmaceutical regimen. Exercise is also part of the U.S. national health guidelines; however, a comprehensive strategy to increase exercise levels is lacking. In the United States, only 23% of adults reported meeting the aerobic and muscle strengthening guidelines in 2019 (75). Overweight and obesity rates have increased throughout the world during the last several decades; COVID-19 shutdowns seem to have exacerbated the problem, particularly among American youth, with overweight and obesity rates increasing from 15.5% in 2018–2019 to 17% in 2021 (76). COVID-19 also increased isolation, a risk factor for depression and other mental health conditions. However, exercise, again, has an important role to play: A recent study in Brazil, for example, found that physical activity during the COVID-19 pandemic helped reduce the risk of depression (27).

Like exercise, changes in diet can induce improvements in a variety of ailments; dietary changes can alter the gut biome leading to salutary systemic changes. Diet and exercise are the touchstones of good health. However, access to both is often shaped by an individual's social context. Those working in lower wage jobs and living in low-income communities have less time, fewer resources, more exposure to environmental stressors, and a larger number of barriers to both healthier foods and regular exercise. The increase in processed food consumption, decrease in exercise, increase in BMI, and higher morbidity related to accompanying health conditions is a worldwide phenomenon. Though the trend is more pronounced in wealthier countries like the United States, it is, nonetheless, occurring everywhere. In the absence of systemic efforts to increase the consumption of healthier foods and exercise, the problems will worsen as will the global reliance on primarily medical or clinical treatments (as opposed to prevention).

The business of health care will boom as excess morbidity and mortality linked to diet and exercise are normalized and accepted as the price of civilization. Health is a function of genes and biology, individual behavior, access to medical care, and social factors called **the social determinants of health**, which are sometimes grouped into five areas: economic stability, education access, health care access, neighborhood and built environment, and community context (29). These factors independently impact health and converge in a myriad of ways. Economic instability is the result of insufficient wages and benefits due to unemployment, underemployment (hourly or part-time work), or contract work. Lower income households have inadequate access to health care, housing, food, and other essentials. Lower income households often make trade-offs between essentials such as health care, housing, and food. These trends are even more pronounced in developing countries. Lower wage jobs carry a higher risk of exposure to harmful work conditions (31). Workers who are required to do physically demanding work for lower pay are often prone to more injuries and ailments (30).

Unemployment is associated with several mental and physical health issues, including high blood pressure, heart disease depression, and anxiety (32). In the United States, many in low-wage or part-time workers lack employer-based health care coverage. Unsurprisingly, health insurance is associated with greater use of clinical services and health monitoring, which is essential for primary prevention (36).

Education is an important determinant of health as it provides a path out of poverty (45). Education provides access to jobs with minimal or no hazard exposure and higher potential for income- and employer-based health care access (33,34). In the United States, children from lower income households are less likely to seek and finish higher education than their higher income peers. Lower income children are often less college ready as their parents have fewer resources to invest in childhood education and development. College education can magnify the advantages of wealth, creating further income inequality. The impact of parental wealth on their children's earning potential is significant in the United States (and Britain), dispelling the myth of meritocracy (46). Education also directly impacts health literacy, including the ability to understand and act on health-promoting behaviors (35, 37).

The importance of education access is underscored by the role higher education plays in creating **economic interconnectedness**, a type of social capital that creates friendships across socioeconomic classes. Recent work by economist Raj Chetty suggests that in the United States economic connectedness is the only type of social capital associated with economic mobility (47). Thus, friendships forged in college between lower and higher income students help propel lower income students up the economic ladder.

Neighborhoods exert a significant impact on both physical and mental health in multiple ways: access to healthy foods, exposure to stressors (e.g., violence, crime, environmental hazards), opportunities for exercise, social capital, and housing. Those living in low-income neighborhoods in the United States often have reduced access to healthy foods and greater exposure to fast foods, convenience, and liquor stores (38). Higher income areas have a greater proportion of full-service restaurants and grocery stores and, consequently, lower rates of obesity, diabetes, and other chronic conditions (39). Lower income neighborhoods in the United States are more likely to suffer from violent crime, which impacts health directly and indirectly. Neighborhoods perceived to be more violent face declining property values, which impacts resident wealth. Additionally, exposure to violence has direct impacts on health and well-being (48). Depression, for example, is directly linked to neighborhood violence (49, 51).

Neighborhoods can also increase exposure to environmental stressors. The environmental justice movement was founded on the premise that all people, irrespective of class or race, should have access to safe, clean, and healthy environments. **Environmental justice** is defined as the "fair treatment and meaningful involvement of all people with respect to development, implementation, and enforcement of environmental laws, regulations and policies" (40). The movement seeks to redress the greater exposure to environmental hazards suffered by poor and minority communities. The U.S. **Environmental Protection Agency (EPA)** describes "fair treatment" as a state in which "no group of people bear a disproportionate share of negative environmental consequences from industrial, governmental, and commercial operations or policies" (40). Ray Bullard, a sociologist by training, is sometimes referred to as the father of the environmental justice moment. His 1990 book, *Dumping in Dixie*, chronicles the history of the environmental justice movement, the disproportionate environmental burdens placed on low-income and minority communities, and ways to advance an environmental justice agenda. Bullard was one of the first to argue that the placement of unwanted land uses, such as landfills or power plants, in low-income or minority communities was denying these communities "equal protection" and is thus a form of discrimination.

Many studies have shown a correlation between demographic variables such as race or income and unwanted land uses or pollution (42, 43). A 2016 study on who lives near the high-risk chemical facilities, as identified by the EPA, revealed that nearly half of those living within a mile of these facilities were minorities (44). The recent Flint water crisis, which resulted in greater exposure to lead and, possibly, harmful bacteria due to pipe corrosion, disproportionately impacted minorities as the city is more than 50% Black. These correlations notwithstanding, the pathways to the disproportionate environmental, and health, burdens faced by low-income and minority communities are complex. In some cases, sites offering the least resistance are often located in low-income and/or minority communities. Sometimes, there is less regulation of sites located in marginalized communities. In other instances, the point source of the hazard (e.g., a polluting factory) is difficult to legally establish without the investment of significant resources, so there is no liability and no compensation for victims. Additionally, research shows that demographic variables such as race or income exert minimal impact on regulatory activity (41). Low-income and minority communities are at greater risk for exposure to environmental hazards and have fewer resources to address the fallout from exposure, including adverse health consequences. They are more likely to be exposed and less likely to have the resources to address the consequences of exposure. In some cases, this creates issues that linger across generations.

The preceding discussion on the social determinants of health shows that health status is the result of a complex interplay between structural and environmental factors over which individuals sometimes have little control and agency, or choices made by the individual. Structural and cultural forces shape, condition, and delimit choices. Even in developed countries, particularly the United States, those who are low income face significant barriers and obstacles to good health. In many ways, the health care system benefits as new clients are created in perpetuity (the same can be argued for the carceral system).

Though much of the discussed research is specific to the United States, the findings are applicable to other nations, especially those with wide chasms between the rich and the poor and without strong social safety nets (i.e., antipoverty programs and policies). In developing countries, these barriers are often amplified as those who are low income are more exploited, have fewer resources and less legal protections, and face greater exposure to violence, conflict, and communicable diseases virtually unknown in richer nations. The next section looks at health disparities, many of which are the result of social factors, in the United States and worldwide.

Health Disparities

Health disparities are created when certain groups experience higher incidence and prevalence of disease, including injuries, due to biological (or genetic), social, or environmental factors. Different morbidity and mortality rates among certain groups from the same health conditions are also examples of health disparities. The reduction in health disparities is an important public health goal as it leads to **health equity**, a state in which no individual is denied the opportunity to realize their full health potential because of their social circumstances (58). The CDC implicates structural factors such as poverty, education, or health care access in the persistence of health disparities. In the United States, disparities exist in all health-related domains, including infectious diseases (e.g., HIV, TB), chronic conditions (e.g., diabetes, hypertension), environmental hazards (e.g., work-related injuries), substance abuse (e.g., alcohol use), health care access (e.g., preventative care and immunizations), and social determinants of health (e.g., education, income, access to healthy foods) (71).

An important disparity in both the United States and developing nations is low birth weight (LBW), a serious condition in which a newborn weighs less than 2,500 grams (5.5 pounds). Low–birth weight babies are disproportionately born to poor mothers in developing countries and Black mothers in the United States. The 2020 LBW incidence for U.S. Black mothers was 14% compared to 6.8% for White mothers and 7.5% for Hispanic mothers. An estimated 20 million newborns annually have low birth weight. South Asia, which includes India, has the largest proportion of LBW newborns (28%) and newborns that are not weighed (66%). Nearly half of the newborns worldwide are not weighed at birth, so the condition's incidence is likely higher. Premature births often result in low birth weight; however, newborns carried to full term who are still low birth weight often have the highest illness risks in the short and long term. Many have cognitive defects and a higher risk for chronic diseases later in life (62). Risk factors include mothers with chronic hypertension in developed countries and mothers who are malnourished in poorer countries.

The differences in life expectancy at birth between wealthy and less wealthy countries, or between the rich and low income in lower income countries, are another important health disparity driven primarily by social determinants such as poverty, health care access, environmental hazard exposure, and nutrition. The differences in disease incidence and prevalence within countries are often stark. For example, HIV, tuberculosis, and malaria collectively cause more than 2 million deaths annually (52). They also tend to impact the marginalized—those who are low income and less educated and who live in rural areas—disproportionately. In sub-Saharan Africa, for example, adolescent girls and young women are at much higher risk of contracting HIV than their male counterparts: Women between 15–24 account for a quarter of the new infections though they are only 10% of the population (52). Those living in urban slums have a five-fold higher risk for TB than the national average, and malaria is more likely to impact those with low income and in agriculture (52). Even in low- and middle- income countries, child mortality rates (< 5) vary significantly by income, mother's education level, and place of residence (urban or rural) (53). Low-income households have nearly double the rate of rich households (90 per 1,000 versus 50 per 1,000). Similarly, as maternal education levels increase, child mortality drops by half (105 per 1,000 versus 55 per 1,000). Rural residents have a child mortality rate that is nearly a third higher than those living in urban areas (80 per 1,000 versus 60 per 1,000) (53).

Gender inequality is implicated in health disparities for a number of chronic and communicable diseases, such as HIV, heart disease, and TB, in both rich and low-income countries (54, 55). Gender inequality can manifest in different ways, including education access, gender-based violence, political participation, and limited health care decision-making. Diphtheria, pertussis, and tetanus (DPT) vaccines are an important part of the immunization regime for young children. The DPT series of vaccines is sometimes used as a proxy for other important childhood immunizations and interaction with the health care system. Those missing all doses of the DPT vaccines often don't have other important immunizations and regular health care access. In a recent study, gender inequality, as measured by the Gender Development Index and Gender Inequality Index, was predictive of low immunization coverage. This suggests that the goal of universal immunization of all children is not possible unless gender-based barriers to health care are addressed (55).

These are just some of the health disparities in the world. The structural causes of disparities are undeniable. The remedies, too often, are clinical or public health based (e.g., health education or immunization campaign), though still very much within the sphere of modern medicine. The impact of clinical medicine and public health measures in reducing incidence, prevalence, morbidity, and mortality are incontrovertible. These achievements notwithstanding, it is important to address the structural roots, or, in the long term, the disparities will continue unabated and may worsen. The social and political will to address structural causes, however, is often low or nonexistent as significant changes have the potential to upend the profitability of the existing system on which so many are feeding. The next section will look at another important structural determinant of health, the urban environment and how land use patterns, specifically in the suburban United States, have caused many of the chronic conditions we see today.

Urbanization and Public Health

Industrialization increased urban populations considerably in a relatively short time frame. New York City, for example, ballooned from half million to 5 million in the 50-year period between 1860 and 1910; in the same time period, Chicago grew from 100,000 to just over 1 million, and Philadelphia's population grew threefold to 1.5 million. The U.S. population also tripled during this time (56). Other industrialized nations experienced similar urban population growth. London, for example, more than doubled its size between 1850 and 1900, growing from 2.8 million to 6.5 million. The pre-industrial world was characterized by high mortality rates due to infectious diseases, which started to drop by the 18th century due to advances in medical knowledge and technologies. For example, Sweden's vital statistics records, which go as far back as the late 17th century, show that population growth was driven primarily by falling all-cause mortality rates. The greater densities associated with early industrialization, poor sanitation and waste infrastructure, poor working conditions, and greater exposure to environmental hazards greatly increased morbidities related to infectious diseases, work injuries, and environmental hazards. Life expectancies in town and cites, not surprisingly, were initially low. In England, for example, urban life expectancies were below the national average for the first half of the 19th century and only started to increase in the late part of the century (57). The urban poor, many of whom were recent migrants from the rural parts of the country, bore the brunt of the health burden as they provided the workforce for the factories and often lived in squalid conditions. This pattern was evident in many early industrializers, such as the United States, France, and Germany, and is seen today in developing countries that are undergoing industrialization and urbanization (57).

Scientific advances, such as germ theory, which replaced *miasma* as the leading explanation for infection, helped provide the scientific basis for public health infrastructure (e. g., water treatment) and programs. In England, the push for change, however, came from increasing enfranchisement during the 19th and early 20th centuries, which eventually gave the vote to all adults of both sexes. Working-class men and women exerted their newfound political power at the ballot box, and politicians responded by building health and social services infrastructure that greatly reduced morbidity and mortality. In 1905, a new national administration in England ushered in a raft of social service programs such as pensions, school meals, school health inspections, sick leave, and unemployment benefits. The English experience was not unique, as various governments at the national and local levels would prove to be important in mitigating the ill effects of industrialization. These experiences offer valuable lessons for developing countries today going through the same growing pains in their march through industrialization.

The section on social determinants of health and the preceding discussion on industrialization and urbanization show that the built environment plays an exceedingly important role in health and well-being. The built environment can facilitate health-promoting behaviors such as walking and bicycling and open space use, foster community ties or social capital, reduce exposure to environmental hazards, provide access to healthier foods, and buffer residents from unsafe or violent events (66). Urban design varies considerably throughout the world, as does the degree to which cities are planned and designed. Some are conceived and developed according to a master plan and others the result of organic growth and ad hoc planning processes.

American visitors to European cities often remark about the low rates of overweight and obesity among Europeans despite a diet that is rich in fats, carbohydrates, and alcohol. These are anecdotal observations, but there is a modicum of truth in them. Europeans have lower overweight and obesity rates than Americans, though the rates have been steadily increasing for some decades, at least partially due to changes in lifestyle (i.e., greater consumption of processed foods, longer working times) due to globalization.[1]

Some of the differences are the result of higher density development in European cities that promotes regular walking and socializing as well as a culture that encourages longer eating times. In contrast, many cities in the United States are suburban and dotted with single-family homes with front and rear yards and two- or three-car garages. As discussed in Chapter 2, the suburban explosion in the United States was driven by many factors, including government policy, the proliferation of automobiles, mortgage finance, and the availability of wood. The single-family home quickly became synonymous with the American dream. The environmental, social, and health costs of low-density, single family–style developments are significant. In comparison to higher density developments, single-family homes utilize more land to house fewer people, which drives up rent and home prices; often contain nonfunctional water-intensive spaces such as grass-filled front yards; utilize more energy and water on a per-capita basis; and make public transportation less cost-effective as there are fewer people per square mile. Single-family developments are also usually auto-centric and discourage walking and bicycling; they promote relatively sedentary and socially isolated lifestyles, which have significant health and social costs. Single family or R1 zoning stipulates many of these requirements, so home developers have little choice but to hew to these design guidelines. Furthermore, existing homeowners are often the biggest NIMBY (not in my backyard) sources protesting any and all new development, especially lower cost, higher density, multifamily homes.

Zoning is a central feature of land use planning in the United States. Zoning originated in Europe as a mechanism for separating residential uses from those (e.g., industrial) with the potential for harming human health. Zoning was quickly adopted in the United States after 1926 when the Supreme Court's ruling in *Euclid v. Ambler* validated zoning as a legitimate **police power**, or power to regulate health, safety, welfare, and morals. Cities in the United States lack a constitutional basis for their police power, so it is devolved, or given, from the states through enabling legislation. Zoning codes stipulate design guidelines such as maximum heights, floor- (built area) to-area (site area) ratios (FAR), setbacks, and permissible or prohibited uses. Some argue that zoning codes have become exceedingly complex and onerous and advocate for a **form–based** approach that focuses primarily on the physical form or external features of the development and less on use.

In the United States, a general or comprehensive plan provides the framework cities use to guide development. Public participation from all community stakeholders is a vital part of the general plan update process and is mandated by law. The general plan should ideally reflect the unique character of each community and its stakeholders' vision for its growth. The plan identifies challenges and contains goals and objectives, within the context of various elements, to address these challenges. The elements are subsections that focus on important community attributes. In California, for example, there are nine required elements: land use, circulation (transportation), housing, conservation, open space, noise, safety, environmental justice, and air quality. General plans touch on sustainabil-

---

1. People are classified as overweight or obese based on their **BMI**, a weight-to-height ratio (weight in kilograms divided by height in meters squared). Typically, a BMI between 18.5–24.9 is considered normal weight, 25 to 29.9 is overweight, and 30 or more is obese.

ity, equity, and health-related concerns in multiple ways and across different elements. For example, many plans attempt to increase walking, bicycling, and public transportation use to both decrease climate change impacts and increase activity levels among community members. The provision of affordable housing is an important component of the housing element, which must identify future affordable housing needs and how these needs will be addressed through future developments. Many cities in California do not encourage higher density development to meet their housing needs. The state government in recent years has become more aggressive in penalizing cities for not meeting their affordable housing needs; time will tell if the state's approach results in more housing being built (for the time being, however, California continues to have some of the highest housing costs in the nation). General plans also identify sources of pollution, including air and noise pollution, and threats to public safety. Some plans address healthy food options and access to employment and education. In California, plans are also required to address disparities in exposure to environmental hazards in the environmental justice element.

Many plans incorporate **new urbanism** ideals that stress higher density developments, walkable and bikeable communities, mixed uses that combine residential with commercial and retail, inclusive communities that provide housing for different demographic groups, public spaces accessible to all members of the community, preservation of natural habitats, and accessible public transportation (64). New urbanist principles seek to reduce per-capita carbon footprint, increase social connections, enable healthier living, and create a higher quality of life for all residents. Research corroborates many of these assumptions as new urban developments have been shown to increase walking, bicycling, diversity, social interaction, and sense of community (65, 66).

Higher density developments, however, are not free from pitfalls. Higher density, mixed-use communities close to public transportation are usually more permeable, or open to outsiders, than lower density communities. Higher permeability can increase the risk for crime. Higher density can also enable rapid transmission of communicable diseases like COVID-19, discussed in the next section.

COVID-19

The epidemiology of COVID-19 and the global response to the pandemic illustrates the intersection of the various dimensions of globalization. The emergence of COVID-19; its subsequent spread; the rapid public health response; disparities in infection and outcomes; and vaccine development and dissemination, as well as the various conspiracy theories; resistance to public health recommendations such as masks, vaccines, and social distancing; and coordinated protests against purported government heavy handedness were all shaped in some way by the forces of urbanization and globalization.

The immediate aftermath of the pandemic provided incontrovertible evidence of anthropogenic environmental impacts. In the days following the mandatory lockdowns or stay-at-home orders, as households isolated themselves and "business as usual" came to a standstill, the smog lifted, air quality improved, and mountains became visible; ecologies healed, and flora and fauna populations rebounded. As the economy slowed, there were other dramatic improvements in the environment, primarily due to reductions in transportation, the biggest contributor to greenhouse gases (GHGs) and other energy-related emissions. Agricultural activity also slowed, further reducing GHGs, soil contaminants, and water pollution. The slowdown in manufacturing and construction activities decreased environmental pressures by reducing resource extraction and pollution. Though much of this slowdown reversed and reached pre-COVID levels when the pandemic waned, the OECD estimates a small but significant (1%–3%) longer term reduction in environmental impacts.

Current theories about the etiology of the virus, specifically its **zoonotic** origins, show that urban encroachment and an increase in animal source foods (ASF) likely played an important role in the evolution and transmission of the virus (69).[2] Zoonotic diseases are those that spread from animal to humans. Most emerging infectious diseases (60%–75%) originate in animals (77). Factory farming practices often pack animals raised for consumption into extremely tight spaces. These conditions increase the speed of disease transmission between animals, thereby raising the probability of a mutation that is infectious to humans. As countries develop, their appetite for **animal source foods** or meat usually increases. The greater meat consumption not only increases risk for new zoonotic diseases but also many of the chronic conditions associated with excessive meat consumption. Additionally, meat production exacts a heavy environmental toll with respect to water and soil pollution and GHG emission.

Even the more controversial claims about the virus's origins have a globalization dimension. These include the theory that United States funded "gain-of-function" research in Wuhan may have been responsible for helping to develop the virus or its precursor (77). These claims show that today the different stages of the research process in all fields, from funding to data dissemination, are truly globalized. Researchers secure funding from entities often not in their home countries, collaborate with those located elsewhere, and share findings in conferences attended by peers from around the world or in papers accessible to anyone with an internet connection. Now, more than ever, we need cooperation across multiple domains if we are to prevent similar outbreaks or other global exigencies (77).

COVID-19 also highlighted many important health disparities. In the United States, for example, the CDC's case surveillance data for age-adjusted infection rates by race and ethnicity through August 2022 show the higher infection toll among some minority communities: Hispanic (21.8 per 100) and Native Hawaiian (22.1) communities had the highest infections rates, followed by American Indian (19.9) and Black (15.6); in contrast, White (14.9) and Asian (11.6) communities had the lowest infection rates. Mortality outcomes also varied considerably by race and ethnicity (67). During the early days of the pandemic, Hispanics and Blacks had a three- to five-fold greater risk of dying than Whites and Asians. The higher risk continued through the Omicron surge in January 2022. Mortality risk fell after the surge for all groups, and the differences by race and ethnicity also narrowed (67).

Several social factors likely contributed to the greater infection and mortality risk among certain minority groups in the United States. SES was an important underlying risk factor, as lower income households tend to have larger household sizes to reduce housing costs, rely on public transportation, and are overrepresented in jobs that require in-person attendance (e.g., construction, food). Health care access is also shaped by income, as many lower income jobs do not provide employer-based health care; lower income households also have less discretionary income to pay for out-of-pocket medical expenses, including testing. The higher rates of overweight and obesity among certain groups also likely contributed to greater COVID-19-related morbidity and mortality as the risk for severe illness and hospitalization increases substantially with obesity (68).

The disparities were worse in developing nations and exacerbated by the economic shutdowns. Those who are low income in developing countries often live in squalid conditions characterized by high population densities and severely inadequate water and waste infrastructure. Additionally, those who are low income often work as day laborers without any social protections such as unemployment, sick leave, or disability. Working conditions expose them daily to multiple hazards. Their infection risk is compounded by malnutrition, which weakens the immune system

---

2. The lab leak theories were initially disregarded as researchers coalesced around the idea that zoonotic jump occurred in a Hunan market, possibly involving pangolins, racoon dogs, and bats. The evidence of a possible inadvertent lab leak, however, has gained more traction recently, though the exact origins will probably remain shrouded in mystery (77).

and increases morbidity and mortality risk. Lack of reliable data from low-income countries makes it difficult to estimate the impact on the most marginalized members of society. The UN estimates that over 100 million people were pushed back into extreme poverty and that food insecurity increased for over 300 million (70). These data suggest that COVID-19 will roll back many of the economic gains made during the last few decades. Additionally, the toll includes noneconomic costs as women's and minority rights, including those of religious and Indigenous communities, have all eroded during the pandemic.

Conclusion

This chapter provides an overview of how globalization has shaped the modern health landscape. In many ways, people are better off as food is more abundant and health technologies have developed treatments for all types of health conditions and extended life spans. The same forces have created an increasing reliance on processed and fast foods, increasing overweight and obesity rates and the chronic conditions, such as diabetes or heart disease, that accompany them. The benefits, however, have been unequally shared given the persistence of health disparities in all nations. The very different experiences of different population groups with COVID-19 brought the social determinants of health front and center. The poor in all countries were more likely to be infected, less likely to receive help, and more likely to suffer long-term health and economic consequences.

Higher income groups, many of whom were able to work from home, saw their investments increase significantly as the stock market rebound and added trillions to their portfolios. They not only escaped the brunt of the virus but became wealthier in the process. The differential impacts played out in many ways. For example, lower income children went to public schools that were unprepared, whereas many private schools adjusted to virtual education more quickly. The costs of the socioemotional and academic damage to lower income children, who were sequestered at home without regular education and peer interaction, will likely impact their mental health and job prospects in the future. Higher income children will not only benefit in the long run from better educational access during the lockdowns, but also the increases in their parent's wealth. COVID-19 worsened inequality across several dimensions, the effects of which will linger for generations.

The forces of globalization also enabled the creation of markets for medical goods and services. This creates significant disincentives for primary prevention, so health conditions are addressed before they occur or when they are in their nascent or early stages through lifestyle and other approaches (e.g., regular health checkups). Medicalization is the preferred approach by globalization as it concentrates power among those licensed to "practice medicine" and legitimates certain treatments, specifically those with a profit potential. Medicalization is as much a feature of modernity, which privileges the scientific method and expert decision-making, as it is of globalization, which turns healing into another commodity in the global marketplace. Though not discussed in the chapter, medicalization also significantly increases resource consumption and waste as treatment and management require more resources and generate more medical waste than prevention. Public health, many argue, should not be driven by markets and profits; health is a fundamental human right. Those working to reduce disparities, however, have little choice but to operate through the channels of the global marketplace where power and resources are concentrated. Much progress has been made, though much more remains to be done, especially in the aftermath of COVID-19, to address health and other disparities (e.g., economic, education, exposure to environmental hazards).

Chapter 9

What Can You Do?

Why Should You Care, Imagining Alternatives, and the Domains of Action

Key Terms

Modernity: The current historical period characterized by the norms, values, and culture that derive from a focus on reason, positivism, empiricism, and the scientific method. Features include scientific and technological growth; an increase in individual freedoms and democracy; rejection of older, nonrational, belief systems; and a reliance on market-based economies.

Civic engagement: Becoming involved in matters outside your household to impact public or collective concerns. Important for the sustenance of democracy.

Morality: Principles used to determine right from wrong. Examples include utilitarianism, Kantian categorical imperatives, and religion (e.g., biblical principles). Moral beliefs are the foundation for the rules and laws used to organize society.

Domains of action: These include the micro (interpersonal), meso (community, city), and macro (national, policy). These distinctions are more conceptual than actual. However, they can help guide where to best focus your energies given your abilities, proclivities (likes and dislikes), and constraints.

Teleology: Simply put, the study of ends and purposes. The belief that in various domains (e.g., nature, economy, etc.) there are specific, desirable, outcomes; the rationality of actions is measured with respect to how likely the actions are to achieve those desirable ends. Democracy and global capitalism are sometimes framed in teleological terms (i.e., these outcomes are desirable and rational and governments must act to achieve them). Teleological thinking can make it difficult to imagine other alternatives, adapt, and change.

Other key terms will be in **bold** and defined in the text.

Review Questions

- What is the "spirit" of sustainability?
- What is the "false dichotomy" with respect to capitalism?
- What is the role of cooperation in the evolution of human society?
- Why is teaching important? How does it subvert the hypothesis that humans are basically "selfish"? Why is mentorship important?
- What are the differences between utilitarianism, libertarianism, and virtue?
- What are the domains of action?
- What are KAPs?
- What is the role of self-assessment in change, and how does one approach this?
- What is vital engagement, and how is it related to "getting involved"?
- Why is art necessary? What is the relationship between art, technology, economy, and sustainability?
- What is "absurdity," and how does it relate to your sustainability efforts?
- Why is it important to separate effort from achievement? How does this relate to sustainability-related actions?
- What is social capital?

Introduction

Sustainability has thus far been defined in terms of the three E's (economic development, environment, equity). These, however, don't quite capture its essence. So, stripped of all formalism, goals, and objectives, what are we really fighting for? Perhaps, it is to let all of creation, all things—alive and not—run their due course and not be hastened to a premature end by our actions. As all things begin and eventually end, their purpose, if there is one, lies in the space in between. It is during their existence, their journey from the beginning to the end, that they sing their song, sharing their spirit, leaving the world richer and fuller in the process. In one sense, all religions are a love song to this spirit—some call it divine—which is both indescribable and ineffable but also palpable and whose presence gives meaning to existence. All aspects of existence are infused with this spirit. At a fundamental level, all things are the same thing, and yet as we arrange and rearrange the building blocks of reality, new things emerge. Some, like the stars, are large, almost immortal, powerful, lifegiving yet lifeless, and others, like bacteria, cannot be seen by the naked eye, and yet grow, thrive, multiply, and perish a thousandfold in the blink of an eye. The sheer diversity of flora and fauna on earth is a testament to life's creativity and richness. The fact that it appears confined to our blue planet, at least in our neck of the galaxy, is all the more reason to protect it with a fervor. Similarly, from the majestic Himalayan peaks, the Iguazu waterfalls, the sandstone layers at Ayers Rock, the aquamarine waters in Polynesia, the world is resplendent with things that are not living, at least not in the way we understand life, and yet they are magical. These too are worthy of fierce stewardship. Ownership to many Native Americans implied such a stewardship, a sacred responsibility to nurture and protect; it did not mean, as we have defined it, the right to use, exclude, dispose, and destroy. Maybe sustainability is recognizing the obligations of stewardship that emerge from our kinship with all of existence; it is the fight to allow all things the space and time to sing their song and share their spirit as long as they can. Rumi, the famous Persian wordsmith, remarked, "*Out beyond ideas of wrongdoing and rightdoing, there is a field. I'll meet you there. When the soul lies down in that grass, the world is too full to talk about.*"

If you are seeking to create a better, more sustainable, world, then know that change starts with you. You must embody that which you wish to create. If you are sincere in your efforts, others will be inspired, and many will follow. You will serve as a mentor, a role model, a teacher without necessarily aiming to be one. The most powerful change agents are those who seek "the good" without an agenda. They act without guile or pretension. This is not a call to sainthood but a reminder that though we are each fallible, we also have the capacity to act, reflect, and improve. It is in the journey—the trying and striving—that we discover ourselves and our power to do good. Dags Hammarskjold,[1] the second UN secretary general, wrote "**Life only demands from you the strength you possess. Only one feat is possible—not to have run away.**" This is the simplest and truest way to phrase what one ought to do at any point: that which one is truly capable of. And, only you can honestly decide that.

The first part of this chapter presents different philosophical approaches to **morality**, specifically utilitarianism, libertarianism, and Aristotelian virtue. The section examines why we should care about sustainability and how we can come together, especially given current societal fissures. The second part urges you to seriously consider that alternatives to the current world order are possible. It argues that there is no particular **teleology** to human evolution; human society is not a simple progression from hunter gatherer to complex, market-based, nation-states. History abounds with a multitude of examples in which societies have chosen alternatives to property ownership, strong governments, money-based market economies, and urban development for specialization and trade. This is crucial, for without this reimagining we are forever locked into making changes within the current "matrix." The last part looks at the different **domains of action,** from the personal to the macro or policy. These are not distinct, and the differences are more heuristic, a learning tool, than actual. The key message in this section is that all change begins with the personal; that by examining oneself—beliefs, capacities, proclivities, and constraints—we can act consciously and purposefully; that by acting consistently in the "spirit" of sustainability, we will enshrine the habits that are necessary and inspire others with our actions; that personal change will effect change in our communities, societies, and the world. These may seem like lofty words or ideas bereft of substance, but they are anything but that. Change, real change, occurs when we consistently reflect, act, and improve, for it is this process that allows us to live the "examined life." The end is the same for all things that exist, but it is in the middle, the journey, that things create and imbue existence with meaning.

Part 1: Why Should I Care?

Some would say that doing our part to save the planet is a moral imperative because every human being benefits from the resources of this planet and the efforts of others. Others may argue that all aspects of the planet, alive and not, have inherent value and should be preserved. Others would assert that since humans are the source of most problems, it is our duty to fix them. The more nihilistic may argue that good and bad are not absolutes and little more than

1. Hammarskjold was widely admired for his unrelenting commitment to peace. The UN's peacekeeping roles expanded considerably during his tenure, sometimes in conflict with the interests of the major world powers. Hammarskjold died in a plane crash, though the plane was likely downed by those opposed to the cease fire he was brokering in the Congo. Hammarskjold's life was a testament to tireless action for the public good. His personal diary, posthumously published as Markings, provides a glimpse into the inner life of a man of faith, who deeply believed in the equality and divinity of all humans, sought to live a life of service to others, and often despaired of his inability to live up to his creed. Markings shows us the inner tumult of a man who outwardly was composed and unwavering. It is a reminder that the best of us are no more free of self-doubt than the rest of us; it serves as a clarion call to action, reflection, and faith that the world is worth fighting for.

the subjective preferences of the majority in a particular society. Thus, what the majority deems desirable becomes "the good." They would not be completely wrong. For example, in the United States, money is the only value that has increased in importance in the last 25 years, with traditional values like community service, patriotism, religion or having children all in marked decline, especially among younger generations. *"There is little in the way of overall consensus about what is most important and what is least important in basic value orientations. American society is characterized by extreme levels of value disagreement rather than broad adherence to a set of common fundamental principles"* (20). If we cannot agree on a fundamental set of values, how can we argue that sustainability is a value that all should embrace? To help provide insight into this question, it is worth considering some important perspectives on morality and how to think about complex moral problems.

Michael Sandel in Justice provides an intellectually rigorous and compelling summary of three different approaches to morality: utilitarianism, liberaltarianism, and virtue (21). Utilitarianism argues that maximizing happiness and minimizing pain should be the goals of individuals and society. Utilitarians would justify taxation of the wealthy as the money can be used to uplift the poor, whose increase in happiness would greatly exceed any pain felt by the rich. Libertarians would reject taxation on the grounds that it amounts to ownership of the individual by the state as a tax is tantamount to the state taking a portion of the individual's labor and hence time. The virtue perspective, sometime called Aristotelian, would argue that it is essential to understand the ultimate purpose, or **telos**, of anything. During Aristotle's time individual taxation did not exist because it was considered an infringement of individual freedom. However, the wealthy, willfully, contributed large sums to public coffers, especially when new public spending was required for the common good. This was called "liturgy." Aristotle viewed the hoarding of wealth as abhorrent and unnatural as wealth becomes a proxy for value. However, it is only in the **polis**, the city (a metaphor for political association), that humans fully realized their telos, as they used language to deliberate and act morally in concert with others. For Aristotle, the act of **civic engagement** was the highest possible virtue.

Sustainability draws on these and other perspectives such as Deep Ecology, which holds that nonhuman aspects of the environment have value apart from their relationship with humans, to extend morality's purview to include other flora and fauna as well as inanimate features of our planet. As with other philosophical questions, there are no easy answers, and even when there is broad consensus on an issue such as poverty reduction (i.e., we must do something about poverty, especially extreme poverty), the potential responses are manifold.

So, why should you be concerned about sustainability, and, if you do care, what can you do to make the world more sustainable? The why is difficult to answer, even if you choose a particular philosophical perspective. To provide some insight, we should perhaps consider what sustainability is fundamentally about, its *telos*.

As alluded to in the introduction, at the level of atoms, the distinction between all things fades. All things that exist in the known universe were contained in the singularity before the big bang, the explosion that resulted in the creation of space-time. In one sense everything that exists today has always existed and will continue to exist. At this scale, creation and destruction are meaningless as things are not destroyed but simply change form. However, at a larger level, the level at which life exists, there is no such guarantee. Life can be extinguished. Species can become extinct. We still do not fully comprehend how a mother's womb converts food into cells that combine to make a fetus, which at some point during gestation becomes conscious and alive. This process is seen as a problem that will be solved scientifically by some and a divine miracle by others. Irrespective of one's beliefs about how life comes to be, consider how unique it is. Though scientists speculate that there are potentially thousands of worlds favorable to life and that sentient life is highly probable in the Milky Way, our home galaxy, as far as we know, life only exists on earth. The sheer scale of the universe, the abundance of stars and planets and the relative absence of life, gives us an inkling of how rare life might be. Even in planets with life, sentient life, or creatures that can think and feel, may be absent. Or, even those worlds with sentient creatures may not have any capability of abstract thinking or the intelligence necessary to produce rockets or art. The rarity and uniqueness of life is a compelling reason to fight for our planet, and not just human life but all flora and fauna with which we share our unique planet. We should certainly aim to become an interstellar species and visit other worlds, perhaps even settle in some; however, the colonization of other planets cannot be our only solution as we will consume, destroy, and discard other places if we do not change our consumptive relationship with our environment. This is the core idea of sustainability: the preservation of our planet's various ecologies so that life in its myriad manifestations can thrive. As the dominant species, and the one most responsible for damaging the planet, it is our moral responsibility—and this can be argued from a utilitarian, libertarian, or a virtue perspective—to minimize anthropogenic environmental damage and limit resource use so future generations of humans and other flora and fauna continue to thrive. Undoubtedly technology will play an important role, but changes in our consumption patterns and lifestyles will have the biggest impact in the years to come.

The previous chapters discussed the evolution of urbanization and trade and provided an overview of macro- and microeconomic concepts, poverty and development theories, demography, environmental issues, and public health. These areas are central to understanding the tension between the different dimensions of sustainability (economy, environment, equity). There is no panacea, or magic bullet, and all solutions entail opportunity costs in other domains.

The economic "E" focuses solely on human well-being, at improving the material conditions for humans; at a bare minimum, economic development seeks to eliminate extreme poverty (i.e., those living below $2.15 per day) and create an economy in which essential needs (i.e., the base of Maslow's pyramid, which includes food, water, and housing) of all humans are met. For the more ambitious, economic development seeks to go beyond basic needs to create conditions in which humans not just survive but thrive, at least economically. The economic goals of every nation aren't that different: Those lower on the development ladder want many of the same goods and services that those in richer economies take as a given.

Globalization has increased material prosperity worldwide and reduced poverty. A much smaller proportion of the world today lives below the extreme poverty line (one in 10) than 50 years ago, though income (i.e., what you make) and wealth (i.e., assets you own) disparities between the rich and the rest are large and growing. Economic development increases female labor force participation, which usually leads to greater social and political equality for women; these gains, however, seem to reduce fertility. As incomes rise and interconnectedness deepens, so do the problems associated with rising prosperity. Health issues across the planet have begun to converge irrespective of development status. Today, chronic conditions related to diet and lifestyle factors are among the biggest causes of premature mortality and morbidity worldwide. Rising living standards also impact the environment negatively by increasing resource consumption, waste production, consumption of animal source foods, and GHGs.

The chapters did not delve into other important ways that globalization intersects with sustainability. The legal system and related inequalities were not discussed, though there are many, nor was the impact on culture. The relationship between globalization and culture is both supportive, or equity promoting, and destructive. Globalization connects the world literally and virtually. Goods and ideas are exchanged through trade, global travel, and the internet. Social media can help organize democratic movements; give voice to the disenfranchised; provide information, training, and knowledge to groups without access to institutions of learning; and provide a global audience for cultural goods (music, art, dance, etc.). Globalization, however, also exerts a strong homogenizing force. In the past, colonialism required physical occupation, but regions today can be culturally colonized without overt violence by global purveyors of culture that export music, food, language, sports, movies, and other cultural goods. Those with established cultural industries, such as Hollywood, or a wealthier population with the ability to produce quality social media content, are more able to export their cultural wares, including their worldviews. The infusion of the same EDM (electronic dance music) beats in pop music worldwide or growth of fast-food restaurants are just some examples of the homogenizing influence of globalization.

The scale and complexity of sustainability-related issues are such that individual action may seem inconsequential at best and ineffectual at worst. This sentiment, while understandable, is incorrect. *At the crux of any sustainability effort is individual action. This simple fact is often lost in the volumes of data and tomes of analysis on sustainability-related issues and solutions. Simply put, you should become more proactive in how you approach your life.* This entails consciously and honestly thinking about your beliefs, attitudes, and actions in the different parts or domains of your life. This will shed light on your motivations and intentions and whether they align with how you wish to see yourself. Similarly, you should scrutinize your relationships, work, recreation, and consumption practices to assess their consistency with your stated values, which, if you are reading this book, includes sustainability. Change is both possible and likely if you regularly engage those areas that are interesting to you and suited to your abilities, temperament, and constraints. Self-assessment, while difficult and sometimes painful, will help you address issues that pique your interest in a way that is, well, sustainable. You may not change the world, but by becoming more conscious of how you lead your life and consistently engaging the issues important to you, you will certainly change yourself and inspire those you interact with. If enough undertake this journey, the world will be better for it. This prescription may seem overly simple, but there are numerous examples of efforts by individuals or small groups that have helped make the world better and more sustainable. These sustainability stalwarts do not possess exceptional talent, ability, or intelligence. However, they punch above their weight because of the consistency of their efforts directed toward a specific arena.

A recent example involving AI is illustrative. Christoph Schuhmann, a computer science and physics teacher in Germany, has helped assemble the largest free repository of images and text descriptions used to train "text-to-image" AI, which can generate novel images based on text entered by users. Using free code from a California-based nonprofit, Common Crawl, Schuhmann and other like-minded enthusiasts, in their spare time, started assembling a database of free images with descriptors. Their intention was to create an open-source, or free, database of images that could be used to train AI and did not belong to any private entity. Today, their database, called LAION (Large Scale Open AI Network) has over 5 billion images and associated text. It is the largest public database of its kind. There are legal, ethical, and other issues with the database as it is not curated or thoroughly vetted; however, its existence and use by large AI companies attests to the capacity of small players to play an outsized role in every arena. Schuhmann and his allies have helped further equity by creating a public database and preventing the monopolization of images needed to train text-to-image AI. This is just one of many examples of the impact of a dedicated group of volunteers.

Involvement or engagement builds **social capital**, a concept that evokes trust, social bonds, cooperation, and collective action. Research across various disciplines demonstrates the vital importance of social capital in addressing a plethora of problems such as crime, unemployment, educational attainment, economic development, and poverty (5). Social capital is also invaluable in helping communities come together during a disaster or a pandemic (6). For these reasons, social capital is sometimes called the glue that binds communities together. Social capital is a function of both formal and informal institutions and relationships. Formal institutions such as governments, banks, and schools permeate modern life. These institutions have rules that are often contractually codified and legally enforced. However, none of these would function effectively without underlying informal codes of conduct, norms, or mores that build and reinforce trust. Effective formal institutions and strong informal bonds enable cooperation, collective problem solving, and inclusivity. Relationships are thus fundamental to building social capital. Relationships can be psychologically satisfying (i.e., provide a sense of belonging and identity) and provide the social connections for more utilitarian purposes such as organizing to accomplish a common goal (e.g., parent–teacher organizations) or personal advancement (e.g., finding a job) (6). Relationships are strengthened through formal and informal civic engagement. Formal engagement includes volunteering with institutions such as the city planning commission or school board, community service–focused organizations (e.g., Kiwanis, Rotary), or fraternal orders (e.g., Elks). Informal engagement with neighbors and others in the community, however, can be more important for creating lasting bonds that allow democracy to thrive. A community in which neighbors socialize and rely on each other (e.g. help with chores, participate in neighborhood watch, etc) and trust in formal govt institutions such as the police are more likely to act collectively for the social good.

When the United States was a fledgling nation in the early 19th century, before it became a military superpower, its strength lay in its high levels of social capital.[2] The U.S. experiment in democracy was keenly observed by Europe. Alexis Tocqueville, a French social scientist, spent 9 months traversing the new nation. His observations, summarized in his groundbreaking book Democracy in America, contained many insights about American society and culture. Notable among these was the American propensity for volunteerism: "In the US, as soon as several inhabitants have taken an opinion or an idea they wish to promote . . . they seek each other out and unite. . . . Americans of all ages, stations in life, all types of dispositions are forever forming associations" (3, 5). Tocqueville opined that this tendency toward civic engagement around issues trivial and serious was perhaps the most important attribute of the nascent democracy.

Volunteerism, however, has declined considerably since Tocqueville. Voting is the easiest way to participate in democracy. It requires choosing between local, state, and national candidates no more than a few times a year. Businesses are required to provide time for in-person voting; mail-in ballots make voting even easier by allowing voting from the comfort of one's home. Yet, Americans are less likely to vote than ever before. In Bowling Alone, Robert Putnam looks at data on voting and civic engagement, using the decline in bowling leagues as a metaphor, to make the case that Americans are slowly disconnecting from the civic, faith-based, social, and other groups that create social capital (22). The use of social media has fed this by channeling people into echo chambers in which they are only surrounded by like and sometimes extreme voices. The result is a deepening isolation and fractured civic space that creates the hyperpolarization seen today.

Once you start engaging, and volunteering, it would be a mistake to expect profound change even if your efforts were profound. What you have control over are your efforts, not the outcomes associated with those efforts. This is not a trivial point as the psychological toll of unfulfilled expectations can lead to cynicism, burnout, discontent, or a complete abandonment of one's goals. If you have resolved to act to make the world more sustainable, approach whatever path you have chosen with the understanding that you are not owed success.

In the Myth of Sisyphus and his other work, Albert Camus suggests that our lives are characterized by absurdity, a quest to seek ultimate answers to life's meaning and our inability to find satisfying answers in any of life's domains, science, art, or otherwise. The absurdity of life is such that a job can become a prison, love can become burdensome, art can lead to despair, and socializing can lead to boredom. The best man can do, Camus suggests, is

---

2. Native American societies often had high levels of social capital as their cultures were more collective with respect to food procurement, defense, or decision-making.

to live, or embrace the "irresolvable meaninglessness." This is another way of saying that, like Sisyphus, you must disconnect effort from outcome: You must push the boulder up the hill, cognizant that it will likely roll back down and you must begin anew. Paradoxically, it is only in accepting this absurdity that you will empower yourself and others to undertake this journey, and what a journey it will be!

You will become less transactional, more compassionate, and without the need for constant self-aggrandizement. Simply put, the perspective that the world exists to serve you will be increasingly replaced by feelings of kinship, love, and connection and not just to others, but to all of existence. The impulse to only serve one's ego resides in all of us and is accentuated by a market-based, capitalist ethos that asks you to only consider your needs, your utility. But, by serving things, causes larger than yourself, by serving others, you will temper these self-focused instincts and, in the process, increasingly accept yourself, and, dare I say, love yourself just as you are and not because of what you have accomplished or accumulated or some other inherent attribute like looks or intelligence.

As discussed earlier, before you can change the world, you must look inward. What you can do today largely depends on what you are willing and able to do. A crucial first step is an assessment of one's motivations, interests, abilities, and constraints. This is a deceptively difficult process as it requires an honest self-appraisal. Don't underestimate the role of neurochemistry, specifically the activation of reward systems in the brain (e.g., dopamine) in response to being socially identified as a "good" person. In other words, sometimes we do good things not only because they are good but because of social rewards such as recognition that sometimes accompany our actions. While it is not wrong to want recognition, if that is the dominant motivation you may unintentionally focus your efforts on activities likely to generate social recognition and, perhaps, not those that maximize your unique abilities.

The self-assessment process can be informal and doesn't necessarily have to involve some sort of psychological survey such as a personality test (e.g., Myers–Briggs, MMPI), though, as will be discussed later, a personal inventory or strengths assessment can be a valuable tool in achieving greater congruence between your skills, interests, inclinations, and work (both paid and volunteer). Together, these create "vital engagement" because you find what you are doing to be enjoyable, useful, and meaningful.

In the beginning, even before you delve deeper into yourself, a simple list of the sustainability-related areas you are interested in will suffice. This list will be the starting point for further research. You can use online research to learn more and then do a deeper dive by engaging those involved in the areas that pique your interest.. At the very least, you will increase your understanding of these topics and may stumble across appropriate opportunities for "getting involved."

You should determine your time constraints so you can devote time regularly, in increments, in a way that makes sense in the context of your other responsibilities. If you choose to volunteer for an existing organization, you may be required to commit a minimum amount of time weekly or monthly; others may only engage you for special events. Your time commitment should be sustainable, which will depend on the other demands on your time. The time you need to rest, recharge, and engage in social and other nourishing activities, including play, should be honestly assessed. Social media and other internet-related activities can be useful in maintaining social networks, staying abreast of current events, and recreation, but many spend far too many hours engaged in online activities. For example, recent surveys suggest that most young adults (age 18–29) use some type of social media, with Instagram (71%), Snapchat (65%), or TikTok (50%) at the top of the list (19). Facebook is the most used site by all age groups, with more than 70% of those between 18–64 reporting use and roughly half of those over 65 (19).

Your skill set will help you decide the type of activities that will be personally satisfying and benefit from your contribution. What you bring to the table will be a combination of your professional and personal skills. Make a comprehensive list of all the things in which you are competent, no matter how seemingly inconsequential or irrelevant the skill to sustainability. For example, you may have experience as a DJ, which may initially seem unrelated to the serious tasks associated with creating a sustainable future; however, the skill will be invaluable in a fundraiser or other community event requiring entertainment. Sometimes the most satisfying activities are those that tap into hobbies that we have, over the years, developed into a veritable skill. By better understanding yourself, you are more likely to choose a program, project, or activity that resonates with who you are and more likely to stick to it long term.

You can engage in action at the micro (personal and interpersonal), meso (local and regional), or macro level (national and global). These distinctions are more conceptual than actual, as many sustainability-related endeavors will cut across all levels. Individual changes in knowledge, attitudes, and behaviors (KAP) will make actions at other levels easier and sustainable. This is another way of saying, "If you wish the change the world, start with yourself." In terms of impact, interpersonal or local actions often yield immediate and discernible impacts. Interpersonal and local action can also be more personally satisfying. For example, helping family or friends or volunteering at a homeless shelter will likely provide some type of direct relief to those you are helping. However, providing assistance one time or consistent volunteering at the shelter may not in any way impact the deeper causes of the problems faced by those close to you or address the structural reasons for homelessness. Conversely, those working on a regional program or a national policy to address homelessness may help reduce its prevalence in the long run, but you may not see any immediate impact from their actions nor derive the same sense of accomplishment as someone providing direct services.

Whatever you choose to do, doing it regularly is the key to self-transformation. Being involved, engaging, becomes a habit, a natural impulse, the more you exercise it. Remember, smaller actions are the necessary precursors to larger ones. You must build your foundation by engaging and acting regularly. And, as you change, you will inspire others by your actions and not just your words.

Sometimes a new journey can be daunting, even for those normally brimming with confidence. Though failure is inevitable and even healthy, setbacks and mistakes, particularly early in the process, can sometimes permanently derail even the most enthusiastic among us. One has to encounter adversity to understand one's response. However, there are ways to make the path more predictable and increase resilience. Mentorship is one such tool. The practice is used in a number of academic and professional arenas. Formally, **mentoring** is defined as "an interpersonal interaction between a seasoned mentor and a novice protégé, which includes supporting, guiding, teaching, encouraging, and role-modeling" (16). No matter who you are and what stage of life you are in, mentorship can help you navigate a new arena by providing both formal and informal guidance and support. Conversely, you can provide mentorship to a mentee if you are seasoned in a particular area. The relationship is mutually beneficial as the mentee gains guidance and support and in return provides feedback and affirmation to the mentor. Later, the mentee may pay it forward by mentoring other newcomers. The American Psychological Association (APA) identifies many different types of mentor–mentee relationships (e.g., teacher–student, professional–professional, peer–peer, friend–friend, etc.), some formal and others less so, and with varying levels of engagement (e.g., daily versus episodic). Role models, even those with whom you don't have an actual relationship, can serve similar roles, particularly if you are unable to find suitable mentors in your professional, academic, and other networks. Identify figures, contemporary or historical, who have done the type of work you are interested in or those who simply inspire you. If your heroes seem mythical, their lives and accomplishments may seem beyond your ken. Find out more about their work, their lives, especially their challenges and how they navigated their path. By finding out more about your role models—the good and the bad—you will humanize them and give yourself permission to make mistakes, learn, and grow.

Part 2: Imagining Alternatives and Leaving the Matrix

In the book's introduction, I mentioned that the complexity of the modern world makes it appear almost magical. It is a world of specialists and interdependencies. The average person has a limited understanding of the technologies, such as computers, or systems, such as banking, that they rely on daily. Even within their domain, the area in which they specialize, there are regions that lie beyond their ken or comprehension. For example, programmers and IT specialists, though working in the same domain, have very different job functions and skill sets. One writes computer code or programs; the other focuses on networking, or communication, between computers. Additionally, neither may have much understanding of the inner workings of hardware, the domain of hardware engineers. This degree of specialization, as shown in previous sections, is essential for creating the economies of scale that allow comparative advantage to flourish and flood our daily lives with an abundance of products. The consequences of this are an increasing material standard of living, which we like, along with increasing environmental and other problems, which we feel guilty about, and a gnawing feeling that we as individuals cannot do anything about these issues. But is this true, and what are some ways of looking at the issues that give us the room to act?

Sustainability seeks to balance the needs of the present with the needs of the future. Not surprisingly, human needs occupy center stage. Though environmental sustainability is an important part of the discussion, it is one of three legs of the sustainability stool. The other two, economic and equity, are more explicitly concerned with human well-being.

Some have made a convincing case for relegating human concerns when considering questions related to sustainability. There are many versions of a nonanthropocentric sustainability paradigm. The more extreme positions call for a culling of the human population as human activities are the prime drivers of environmental degradation and climate change. Others argue for a radical change in the utilitarian manner in which humans approach nature, as a thing to be controlled, mastered, and transmuted into an economic good. This perspective requires a decisive shift away from the dominant economic system today, global capitalism. Other critiques argue that private property, the heart of capitalism, is largely inconsistent with true equality and freedom. The data on increasing income and wealth inequality would certainly lend credence to this view. The critique, however, is deeper than dollars and cents. It holds that humans cannot be free from coercion by others and systems of power (a la Foucault) as long as the primary driving force is acquisition and the hoarding of private property. One can only be dispossessed of a thing if one truly owned that thing. Property rights are, however, the sine qua non (essential element) of global capitalism. They are the bedrock of modern society. Though democracy promises freedom, our actions are significantly constrained by property rights that inhere in all goods and services. Sustainability, as has been shown throughout this book, is inextricably linked to property rights. Can we conceive a world—and there is ample historical precedent—without property rights? What would such a world look like? What types of incentives would exist for work? In our world, carrots and sticks compel people to put in the excess hours needed to produce many of the products of the modern world. Property ownership and its attendant benefits are the carrot, and coercion, whether it's legal or implied, such as the threat of losing one's job, is the stick.

A related question is "How does one conceptualize progress?" Is it measured primarily in terms of scientific and technological growth, the degree to which technology permeates society and daily life? The modern world is certainly a technological marvel. Machines of all kinds suffuse every aspect of our daily lives, in some instances replacing human labor and augmenting human capacities in others. Machines have enabled economies of scale, allowing for the production of higher quality, abundant, and cheaper consumer products. AI is also increasingly competent at higher level, creative tasks. To some, machine intelligence poses an existential threat to humanity as, at the very least, it will obviate human labor in many fields, increasing un- and underemployment, or, in a doomsday-type scenario, take punitive actions against humans. To others, AI will improve human endeavors in every arena and further free people from mundane tasks.

Is progress synonymous with certain values? Some, like Steven Pinker, suggest that human sensibilities have evolved considerably since the enlightenment with the adoption of reason, democracy, and some framework of fundamental human rights by most societies across the planet. Market-based global capitalism that promotes interdependency through trade is often seen as a necessary component of this evolution. Others, like John Gray, suggest that the relative calm of the past century after WWII is a lull before the inevitable storm; humans are mostly unchanged and now have even deadlier tools to engage in war. The Russian–Ukraine conflict and the expansion of NATO; the Israeli-Hamas conflict; the simmering geopolitical tensions between many nuclear armed states such as the United States, China, India, and Pakistan; the resurgence of the Taliban; and the rise of right-wing politicians across the globe lend credence to those skeptical of human social progress.

It is worth considering that the central conceit of **modernity**, the apotheosis of enlightenment values and reason, is its Achilles heel. Modernity sacrifices all other ontologies, or ways of being, and views them as primitive or in transition, at the altar of scientific progress and individualism. That is, the progress of every society is measured with the yardstick of how far along it is in the democratic-global-capitalist path. In our adherence to this teleology, that only one type of societal evolution is correct and desirable, we rob ourselves of the capacity to imagine other worlds. So deep is this belief in the essential correctness of modernity, and by extension global capitalism, that radical critiques, those that call into question the very assumptions of the system, are cast as loony and only those who seek to provide a salve or a fix without altering the system in any substantive way are given wide audience.

What gives this system its power to resist change? Undoubtedly, the dramatic increase in the material conditions of life since the industrial revolution provides a powerful rebuke to anyone suggesting that other systems, other ontologies, can do better. If better is defined primarily in terms of material things that make life easier and more pleasant, the battle is perhaps lost. Free market disciples recount the story of Boris Yeltsin, who wasn't yet the Russian prime minister, visiting a Texas grocery store and remarking that even the Russian elite didn't have access to such abundant food as a cautionary tale about pitfalls of socialism or communism. However, the effects of producing this abundance on the health of the planet and humans is elided (e.g., Americans have among the highest overweight and obesity rates in the world and spend more per capita on health care than any other country). Furthermore, by setting up a choice only between capitalism and socialism (i.e., the proverbial straw man), alternative ways of organizing production and economies aren't considered (sometimes this is referred to as a "**false dichotomy**").

However, there is a modicum of hope if we are willing to at least entertain the idea that material and technological progress, though astounding and appealing in many ways, are also deleterious. The earlier chapters in this book have provided ample evidence of the destruction wreaked by global capitalism on the environment. The impact on public health, as discussed in Chapter 8, is similarly enormous. Science and technology have certainly enabled a better understanding of diseases and improved clinical treatment. However, the root causes of the biggest threats to human health (heart disease, cancers, etc.) are lifestyle and environmental factors created and sustained by the global capitalist system. *The system creates problems that it then attempts to fix exclusively through new modalities and markets, and the game continues. By creating both problems and responses to these problems, the system increasingly legitimizes itself as the only real thing. Those living within it are unable or unwilling to imagine other worlds, and when they do, these worlds seem inhospitable, undesirable, retrograde (backward), and even inhuman.*

A truly expansive view of sustainability must consider societal forms that do not adhere to dominant narratives about progress. We must seriously examine different ways to live and the promises they hold for not only human well-being but also the planet. Every action has costs. Some are direct and easy to discern, others are indirect, sometimes called opportunity costs, and harder to determine. Our inability or unwillingness to imagine other ontologies, ways of being, is perhaps the biggest impediment to change.

Arguments that valorize self-interest, individual utility, and other similar concepts (e.g., the selfish gene) are usually based on the idea that humans are hard-wired to select for these traits; nature has selected for certain genes that create certain proclivities, self-interest being at the top of that list. These beliefs are the basis of modern economics and used as a powerful rejoinder against arguments that suggest more utopian or communitarian ideals. Markets are the solution to the problems created by self-interest. Free, fair, and open markets ensure that all voices are heard and optimal solutions emerge. However, even remotely free or fair markets have never existed as they are hampered by the same traits that requires their creation in the first place, namely the human tendency to act selfishly to subvert the ideals of the market, by lying, cheating, and stealing, whenever possible. Instead of recognizing the impossibility of perfection given the very human traits that ostensibly require markets, proponents of markets merely double down on their claims and argue for improvements in social and political factors that will allow true markets to emerge.

There is no doubt that self-interest plays a powerful role in human evolution, but it is not the only natural impulse and perhaps not the most important one (it is, however, the easiest of our default states, though in the long run doing only things in our immediate self-interest is often not in our individual or collective self-interest). Additionally, even if the "selfish" hypothesis is somewhat true, what do we know about our capacity to create new culture, new ways of being (ontologies), that are not grounded in purely self-interest?*Perhaps a better approach might be to recognize the limits of a society based on only one or a few related traits such as selfishness, individual utility, altruism, or communalism.*

What does human history teach us about how humans are hardwired? In Blueprint, Nicholas Christakis elegantly lays out the historical evidence for what he terms "**the social suite**," a set of characteristics present in all societies (24). These include individual and community-building traits such as recognition of individual identity, love, friendship, social networks, cooperation, and altruism, as well as those that foster social separation (in-group bias, hierarchy) (24). Learning through observation is promoted by the social suite and helps create culture as knowledge transfer is much easier in a group setting than a solitary one. One can learn complex tasks or avoid costly mistakes by simply observing others in the group. Teaching is also part of the social suite.

Teaching increases learning efficiency, but it only occurs in a few species besides humans, as it exacts too high a cost for the teacher, who may gain nothing in return. If we are purely selfish, then why do we teach? Because teaching is the bedrock of human society. Through teaching, formal and informal, we transmit skills, knowledge, and culture, including morality. Our first teachers are our parents and others in our social circle who teach us how to survive, communicate, and socialize. As we mature, other teachers provide us with the tools to become productive members of society, contributing to the sustenance and development of human civilization. There is no one way to teach as there is no one way to learn. We learn through one or more of our senses, through emulation, abstract methods (writing), repetition, and feedback. Many of our greatest teachers have been those whose lessons are found in their lives. The list is long and includes some such as Jesus or Buddha who have become more than human and others, such as Nelson Mandela, Thoreau, or Gandhi, whose lives are a testament to a radical commitment to their values, often at great personal cost. Good teachers, famous or not, not only inculcate utilitarian skills but also inspire us and help develop our moral compass. They provide us with the tools to examine ourselves, others, and society. They help us believe in ourselves, even in the darkest of times, so we continue to grow in all facets of our lives long after we have left their tutelage.

David Graeber and David Wengrow in The Dawn of Everything: A New History of Humanity provide abundant anthropological and archaeological evidence for the different ways humans have organized themselves (23). Their sweeping survey shatters prevailing myths about the development of agriculture, urbanization, political and social hierarchy, private property, democracy, and technology. They show that technological growth sometimes serves only an artistic function as during pre-Neolithic times when ceramics were used for figurines and not for storage and cooking; technology wasn't always used for commerce or warfare, as when the Greeks used the principles of steam power to open temple doors and similar theatrical displays or when the Chinese used gun powder for fireworks; people experimented with agriculture for thousands of years, sometimes switching back to foraging; agriculture did not always lead to private land ownership and sometimes, as in the Fertile Crescent, led to the opposite; cities were created for a number of reasons, including as ceremonial centers as in Mesoamerica or Poverty Point in North America, and often without a strong hierarchy or wealth concentration as in the highly urbanized cities in Indus Valley. The sheer diversity in social and political forms are also chronicled, perhaps none as humorous and incisive as those by Native Americans in their rejection of European notions about private property and the resulting social and political hierarchies. Kandiaronk, a Wendat statesman, noted, "If you abandoned conceptions of mine and thine, yes, such distinctions between men would dissolve, a levelling equality would then take place . . . and yes, for the first 30 years, you would see a desolation among those only qualified to eat, drink and sleep . . . but their progeny would be fit for our way of living." His words ring truer than ever today as collectively we are more dependent on systems of production than ourselves for all needs, basic and otherwise.

There are many more examples that upturn our "modern" conventional understanding of human history and progress. Our modern global capitalist society is simply one arrangement and may have run its course given our increasing subjugation to systems of control in every arena of our lives (e.g., the rise of specialists in every area and their state-sanctioned ability, licenses, to determine what is correct) and the enormous environmental havoc created by our modes of production (e.g., specialization, industrial production relying on economies of scale and trade) and institutions of finance (e.g., investment / retail banks and related institutions) that rely on them. If we are to imagine different futures and experiment with alternatives, we must first believe that there is no teleology to human society, that alternatives are both possible and desirable, and we can act to create new realities.

Perhaps the AI revolution will offer us one pathway to a better world. Much has been made of the AI threat to human jobs along with other existential fears. In many jobs, humans will increasingly be replaced by machines. Large-scale unemployment and underemployment are certainly concerns that must be addressed by policymakers. However, AI can also free us to experiment with how we live if basic necessities are provided by machines. Some view this as disastrous. They imagine a world of mass unemployment, poverty, despair and related ills (e.g., substance abuse, crime, mental health, and other problems). However, these fears only make sense in the context of our modern economy. Many today work in jobs that don't provide enough time and money for basic needs and/or discretionary spending on leisure, personal growth, and the like.

It's also possible to create a world in which humans freed of the need to work in meaningless jobs are able to thrive, much as many Native Americans did prior to the arrival of Europeans. However, to achieve such a society, *we must first leave behind the nostrum that all roads culminate in a global capitalist society and that all problems can only be solved through its framework.*

But What About Art?

The Renaissance, which means rebirth or renewal, was an extraordinary period of artistic creation in Italy. Many attribute the "rebirth" to financial support provided to artists by bankers, financiers, and traders such as the Medici family. Similar arguments are deployed when discussing the art "economy" today. However, such arguments fundamentally misunderstand why humans create art, which predates language, agriculture, or economy. It has existed since our earliest ancestors created jewelry and pottery, drew on cave walls, or made bone flutes. It has existed since we played our first games as play and sports are also part of our artistic canon. Artistic expression is as varied as the individuals that comprise humanity. Through art we understand what it means to be alive. It is essential for our minds, our psyche, as much as eating and drinking are needed for the physical body. Capitalism has provided markets for art and allowed some artists to attain fame and fortune. It is no surprise that artists in every field discuss failure and success in terms of the business of their art (e.g., the music business). However, for every popular or wealthy artist there are scores others who are neither famous nor rich but produce art that is appreciated by none or a few or that, perhaps, will be lauded posthumously.

One could also ask if the modern economy has resulted in only a few pursuing art seriously as the economic opportunity costs are significant. The free market proponents would argue that only the best and most committed will survive, and this is desirable as only the best art is produced. However, this is simply not the case. Any serious musician or other artist will quickly confirm that art produced for the masses (e.g., pop music or art) is often not complex or innovative. For example, the use of auto-tune by singers, overproduced songs in which all mistakes have been removed through digital editing, generic melodies, and a limited compendium of beats are all characteristic of modern pop music. Additionally, in the age of social media, marketing plays an outsized role in promoting certain types of art at the cost of others that are more difficult to understand and process. But there are bright spots. Amateur artists in all disciplines have access to a variety of tools to learn, improve, and market their art. Online tutorials and lessons, software, and social media provide unprecedented opportunities for those without the resources, time, or talent to still engage art at a very high level. In many ways, the capacity of dilettantes to pursue their art is greater than ever before.

One could also make the case that we have an excess of art, particularly art geared toward mass consumption. Those in the business collect and analyze reams of data to fine-tune the "formula" for music, films, television, and the like. The resulting deluge of art on Netflix or released by top-selling artists is not so much the artistic expression of an individual or group but more a vehicle to garner enough mass attention to justify advertising fees of those subsidizing the art.

The political economy of art sees it as any other consumer object subject to market forces. Many who buy art do it to diversify their investment portfolio. Some buy art for private collections, for pleasure, and as an investment, but sometimes they take it out of the public realm. However, art, at its core, has little to do with its role as a consumer good. It can and has existed without markets. It exists to create meaning, to help make sense of the world, to help make sense of being alive while slowly dying. Art helps us grapple with existence, its simultaneous meaninglessness and profundity. *Sustainability and art are intimately linked* as a more sustainable world allows the artist and the muse to exist as long as they must, engaged in a dance, sometimes exchanging places with a muse, becoming the artist, and vice versa, each helping the other understand what it means to exist.

Part 3: Taking Action

Micro: Personal

The KAP paradigm is sometimes used in social science to understand behavior change. Knowledge is awareness, attitude is orientation, and practice is behavior. It is, of course, entirely possible to engage in certain behaviors without having an understanding (knowledge) or predisposition (attitude) to act. We often do things without a clear understanding of why we are doing those things. Sometimes we engage in certain actions to "fit in" or because the behavior is socially desirable. We act to gain the approval of others. Social media is rife with content designed to generate "likes," "hearts," and other symbols of social approval. This generates a small dopamine surge, a response in our brain's pleasure centers. The positive feelings are ephemeral, short-lived, and must be renewed with new postings, continuing the cycle. The underlying motivation has less to do with the value of the action captured in social media and more about the constant, fleeting, feelings of pleasure generated by the "likes."

If you want to persist, particularly when you're going against the grain or contrary to popular opinion, then knowing why you're doing something is important. It is necessary to separate the needs of your ego from the value of the action you are undertaking. This is not to say that you shouldn't accept or value social approval; always be grateful for the positive affirmations and good will of others. Rather, what others may or may not say in response to what you are doing should not be the primary motivation. You should be able to persist without constant social approval and, sometimes, the passion or zeal that motivated you earlier in the journey. To keep going, you better have good reasons for why you chose this journey in the first place.

Lasting personal change can be difficult to achieve. "People never change" is an oft-repeated mantra, but what is the evidence for this shibboleth, this widely held belief? What are, if any, the personality characteristics that remain invariant over a lifetime? Psychologists suggest that neuroticism, extraversion, openness to experience, agreeableness, and conscientiousness are the biggest core personality traits. Some researchers have argued that these persist throughout adulthood. Others suggest that the stability of these traits depends largely on external circumstances (10, 11). In other words, you may have certain predispositions that are hardwired at birth, but your circumstances and life stage will help drive how you respond. These "characteristic adaptations" to your reality include your goals, coping skills, and values and beliefs within the context of your life stage (10). Furthermore, your goals will usually be concerned with one of the following: work, achievement, relationships, meaning (spirituality), and generativity (societal contribution, legacy). Research suggests that those who focus primarily on work and achievement are less happy than those who pursue goals related to relationships, meaning, or generativity (10, 12). By looking at ways to make the world more sustainable, you are pivoting toward goals that provide meaning, improve relationships, and leave a positive legacy. A byproduct of this orientation is likely to be greater personal well-being and happiness.

Change can be difficult and painful. Often change is thrust upon us by circumstances outside our control, by trauma and adversity. By meeting new challenges, especially traumatic events, we discover our ability to persevere or endure. This gives us confidence in our capacity to meet other challenges, which can also help strengthen our relationships as we appreciate those who are in our lives, particularly those who aid us, in trying times. Adversity can also help orient us toward the present (i.e., be in the moment) as we realize the future is not guaranteed. Adversity can provide clarity. People are indeed more receptive to change when faced with traumatic or adverse events as we reevaluate our lives and our goals. However, the window can close quickly. In the absence of concrete steps taken toward forming new habits or practices, over time, we often revert to how we were.

Can you achieve some congruence between your core personality traits and your adaptations (goals, relationships, generativity) in the absence of trauma or adversity? Can you write a life story that melds who you are with your circumstances to realize what you want to become? Can you create challenging situations that enable growth? The short answer is yes. However, this process of "**vital engagement**" requires a deeper dive into yourself and your environment. It is through this process that you can start to develop a deep, engaged, and meaningful relationship between yourself and the many domains of your life, including your interests in sustainability. Vital engagement is not simply a function of your personality or your environment, but rather how you mix and match the two to derive meaning and become socially useful (13, 14). Vital engagement is different than work engagement. People can maintain short-term engagement with their work, even if its inconsistent with their values, if it provides recognition, stimulation, and monetary reward. However, long term engagement becomes difficult if the work is not meaningful at a deep level (14). Identifying, understanding, and then leveraging one's strengths is an important component of vital engagement. Strengths are ways of thinking, behaving, and feeling that feel natural and not contrived; strengths energize individuals (15). You are better adjusted and happier when there is congruence between your biophysical traits (your core traits including your strengths), your adaptation to your circumstances (your psychological response to your environment), and your larger life story (the role you play in society). It all starts with understanding who you are, what you want to do, what you can do given where you are at this moment in your life, and how all of that fits into your larger life story. Introspection, conversations with others, personality, strength or aptitude tests, and formal therapy can all help elucidate these. You can also simply start with writing down who you think you are with respect to your goals, likes and dislikes, strengths and weaknesses, and the components of a successful life. Additionally, as discussed earlier, write down the areas in sustainability that are interesting and important to you and why. This list will be the beginning of your journey to achieve some consistency between who you are, what you can do, and what you want to do to help make the world more sustainable. As you gain further insight into yourself, your life, and the areas you have chosen to impact, you will refine your ideas. This process does not have to be formal. You can write down your ideas and experiences or keep an audio or video record. Whatever you choose to do, a record can be invaluable as it will help remind you of your good and bad ideas, the things that have worked and those that haven't. It will provide perspective as you see what you were and what you have become; it will enable compassion, as we are more likely to give grace when we see our faults and the grace extended to us by others. If you choose not to record your thoughts, at the very least, regularly engage in introspection to consciously remember the important parts of your journey. But always remember, this is a journey, a process and not a destination.

Micro: Interpersonal

The interpersonal sphere encompasses your family and friends. During the early days of the pandemic, over half (51%) of the U.S. population over 16 reported helping their neighbors informally. Nearly a quarter in the same age group volunteered formally (1). Action in the interpersonal sphere can be something simple, such as helping a neighbor with a household chore or assuming a greater share of responsibilities around the home. It's important to cultivate a disposition, a way of being, whose default mode is to consider another's perspective and be helpful. This can be difficult as we often tend to think in terms of personal convenience, our rights and contributions. These default states can lead to feelings of "being wronged," which make compromise and compassion difficult. Among married couples, for example, each spouse tends to claim a larger proportion of housework (2). Researchers have documented similar findings in different group settings, with people frequently overestimating their contribution to the social good. We spend a lot of resources on fighting for our rights, though in many instances, such as divorce proceedings or accident liability, culpability is shared, and each litigating party bears some of the blame. Even when individuals are taught to identify biases, they rarely apply the same scrutiny to themselves though they will happily point out the faults of others (2). Researchers suggest that this is because we believe in our own objectivity: We see the world as it is, and others are not quite capable of that. An important step in developing a positive, open, and helpful interpersonal orientation is the cultivation of *a healthy skepticism about one's ability to completely understand the world and the relative value of one's actions*. Being open to others also means being open to being wrong about how we see the world. This subtle but more important change, if practiced diligently, will help reduce arguments, foster cooperation, and make you more willing to help others even when you feel you are doing more than your fair share.

Meso: Local and Regional

An important initial step in getting involved is developing a better understanding of civic affairs locally. You can start by researching your local government and related entities (e.g., city council, school board, planning commission, citizen advisory groups, etc.) to understand their organization and ambit (sphere of action and influence). This process will help you understand where and how you can get involved locally to impact public policy. In the United States, citizen input is often sought by government for important planning and policy matters. Local governments in California, for example, are required to seek citizen input when updating the city general plan, which includes policies governing land use, housing, transportation, resource conservation, open space, public safety, noise, air quality, and environmental justice. Similarly, many development decisions at the city level require public hearings. In both instances, a wide swath of resident voices is often absent as many residents simply don't engage local affairs even when decisions taken by local government are likely to impact them. Input is often given by those with vested interests in the matters being considered and other "regulars" at the meetings.

The distinction between local and regional depends on the political or administrative subdivisions in a country. Usually, cities or towns in most countries have their own governmental apparatus tasked with managing municipal concerns, which can include land use, public safety, infrastructure, and schools. In the United States, there are usually four levels of government: municipal (city), regional (county), state, and federal (central). The constitution of the United States, however, recognizes only state and federal governments. During the past 100 or so years, the police power, or power to regulate at the local or municipal level, which includes zoning, has been given by the states to cities through enabling legislation and was recognized by the Supreme Court (e.g., the 1926 case, Euclid v. Ambler Realty, which validated zoning as valid exercise of local police power). Similarly, state law governs the structure and powers of a county, which include the relationship between counties and municipal governments. In California, for example, county governments provide health and social services, collect property tax, record legal documents (births, deaths, marriage), register voters, administer elections, conduct regional planning, and manage essential infrastructure, including water, energy, and transportation.

County governments and related entities often engage in sustainability-related efforts through one or more of their offices. For example, the Southern California Association of Governments (SCAG), a metropolitan planning organization serving six counties, produces a regional plan called Connect SoCal that combines transportation, housing, development, and environment to meet state and federal infrastructure and sustainability goals. The development of a multimodal transportation system that reduces single-person automobile use by increasing bus, light rail, and bicycle use and promoting walking is a key component of SCAG's transportation plan as any reduction in congestion will impact climate change and public health. Similarly, any increase in bicycling and walking will improve public health as many Americans don't exercise regularly (consequently, overweight and obesity rates are high and increasing along with their attendant chronic diseases).

If you are interested in any of these sustainability issues or others, there are many ways, formal and informal, you can become involved locally and regionally. Your neighborhood and surrounding community are your immediate local sphere. This is the easiest of the higher levels to engage and offers a variety of opportunities that do not require specialized skills. Furthermore, the scale is small enough that you will often be able to witness, firsthand, the impact of your efforts.

If you don't want to become involved with local government, at the very least you can become a better neighbor. This may mean watching your neighbor's kids, sharing food, helping neighbors, especially those with disabilities or other limitations, with chores, or joining the neighborhood watch. There are many different actions you can take in your own neighborhood, but the first step is stepping outside the comfort of your home to engage those in your immediate vicinity.

If you want to make volunteering a habit, join a local group that resonates with your interests. There are many local groups that have a long history of serving the public good, most of which have seen declining membership in the last few decades. There are many faith-based organizations with varying degrees of community or external focus. Even those focused mostly inward, on their members, help foster spirituality, social capital, and a sense of community. Some faith-based organizations regularly provide needed services for the larger community. Their impact extends beyond their immediate membership. You can participate in these efforts even if you are not a member. A church in my community, for example, provides regular showers for the homeless, through a mobile shower van, and washes their clothes. Local volunteers, many who are not members of the church, help wash clothes and perform other essential tasks. A local nonfaith-based community group to which I belong regularly donates money to help defray some of the costs associated with the mobile shower. This is a wonderful example of different groups and private individuals coming together to address a need in the community. There are many such examples in communities throughout the world. If you look deeply enough, you will find a group or an effort that resonates with you. In addition to local faith-based institutions, you can explore one of the many fraternal and service groups (e.g., Kiwanis, Rotary, Soroptimist), civic organizations (e.g., ACLU, NOW), nonprofit or nongovernment organizations (e.g., OXFAM, Red Cross), neighborhood groups (e.g., HOA, Neighborhood Watch), or government- or education-related groups (e.g., volunteer firefighter, planning commission, parent–teacher organization). Groups are sometimes organized by area of focus (e.g., environment, poverty, health, education), demography (e.g., youth, elderly, women), geography (e.g., community, city, region) or other characteristic (e.g., Alcoholics Anonymous). There is no shortage of groups or paths you can take once you have a better sense of your interests, abilities, and constraints. Some may require formal credentials, but many ask for nothing except regular engagement. An important personal benefit is the kinship you will develop with others in the group. Some may become lifelong friends as the bonds you develop will be based on a common purpose.

Macro: Global

Macro-level involvement usually means engaging policies or programs with a wide sphere of influence. This often means engaging some type of government or related entity.

At the national level, the Environmental Protection Agency (EPA) is the steward for U.S. environmental policy and programs. Other important national agencies include the National Oceanic and Atmospheric Administration (NOAA) in the Department of Commerce; the Department of the Interior, which includes the Bureau of Land Management and the Geological Survey; and the Department of Agriculture, which includes the U.S. Forest Service. There are many other national agencies, many with overlapping areas of influence, tasked with sustainability-related goals. Many states have similar agencies, some that go beyond the requirements or mandates imposed by federal agencies. For example, the Governor's Office for Planning and Research (OPR) in California administers the California Environmental Quality Act (CEQA) as well as policies and programs related to economic development. It is a "one-stop shop" for sustainability-related issues.

If you want to work on sustainability issues at the global level, there are many intergovernmental organizations that are regionally specific (European Union, Association of South East Asian Nations, African Union) or topic specific (OECD, G7, or G20). Additionally, there are many nongovernment organizations (NGOs) organized around a specific area such as health (Doctors without Borders), international aid and development (Oxfam, Red Cross), or conservation (World Wildlife Fund). The largest intergovernmental organization is the United Nations, which has many departments and programs focused on one or more dimensions of sustainability. These include organizations focused on economic stability, development, and trade (IMF, World Bank), health (the WHO), children (UNICEF), and food (FAO). There are many other programs and offices at the UN aimed at peacekeeping, conflict resolution, humanitarian efforts, and scientific advancement. The UN was created to provide a forum for countries to address conflicts and deepen ties. Its structure and workings have often been heavily criticized, but it is presently the largest intergovernmental organization in the world. Its SDG framework, discussed in Chapter 2, articulates goals and objectives in 17 related areas that member nations have agreed to pursue to address climate change, protect the environment and resources, increase economic prosperity, promote cooperation, and create a more equitable world for historically marginalized groups.

The UN has helped broker multilateral climate change agreements such as the Kyoto Protocol and Paris Accords. The UN also currently convenes the largest research body tasked with summarizing the state of "scientific, technical and socio-economic knowledge on climate change, its impact, future risks and options for reducing the rate of climate change" (17). The UN's Intergovernmental Panel on Climate Change (IPCC) is comprised of experts from every field who review and summarize the latest research on climate change, including likely impacts; identify areas where knowledge is lacking; and provide options for "adaptation and mitigation" to address these challenges. The IPCC assessment reports are, however, "policy neutral" in that they provide actionable information to policymakers but do not recommend specific actions (17).

Conclusion

Sustainability is not simply a tool, technique, skill, or policy. It is the belief that this world and everything in it is intrinsically valuable; that there is flow to creation and destruction whose equilibrium human civilization has tilted heavily in its favor. This tilt is often not for any human life-sustaining functions; in fact, in some cases, the (over)use of resources is to the detriment of human welfare—but serves a political economy, global capitalism, to which all are subject and which increasingly operates simply to sustain itself. ***The machine has become the meaning.*** Typical responses to this dilemma include "green" technologies, such as solar, cold fusion, electric vehicles, Galvorn (a new material), or green policies and programs, such as Pro Environmental Behaviors (PEB) or LEED. These often act like bandages instead of compelling introspection into why we need so much energy or other resources. They prevent our evolution from utility maximizing agents operating in various markets. Each technology or program also engenders additional issues, and we are forever locked into searching for yet newer technologies or implementing newer programs or polices. We become the patient who needs a regimen of more and more pills to address chronic health problems mostly caused by overconsumption and lack of movement and other pills to diminish the unwanted effects of the first set of pills. The machine—global capitalism—however, likes this process as it continually creates new markets and new ways to generate profit. We benefit from the profit and new technologies, but increase the alienation from ourselves, each other, and our environment. Though much can be accomplished by tweaking the machine, and we should work towards that, fundamental change will not come from within the machine. For this, we must dive deep into ourselves, to lose the beliefs that shackle us, keeping us tied to the machine because we fear failure, discomfort, ineptitude, or mockery. We must imagine ourselves as capable of more and then take concrete, tangible steps toward that "more," toward that different world. There is no guarantee that we will accomplish anything substantial, but that is not the point. In the end, we will be changed, and in the process so will other aspects of the world with which we collide.

References

Chapter 1: The Big Ideas: Globalization, Urbanization and Sustainability

1. *Monticello. (n.d.). Wealth of Nations.* https://www.monticello.org/research-education/thomas-jefferson-encyclopedia/wealth-nations/

2. History.com Editors. (2019 May 2). *Model T.* https://www.history.com/topics/inventions/model-t

3. Wray I. (2019). *No little plans.* Routledge.

4. OpenStax. (n.d.). 33.1 *absolute and comparative advantage.* Principles of Economics. https://openstax.org/books/principles-economics/pages/33-1-absolute-and-comparative-advantage

5. Skousen, M. (2007). *The big three in economics.* ME Sharpe.

6. OpenStax. (n.d.). *Introduction to perfect competition.* Principles of Economics. https://openstax.org/books/principles-economics/pages/8-introduction-to-perfect-competition

7. Jahan, S., Mahmud, A. S., & Papageorgiou, C. (2014, September). *What is Keynesian economics?* International Monetary Fund. https://www.imf.org/external/pubs/ft/fandd/2014/09/basics.htm

8. Skousen (2007).

9. Environmental Protection Agency. (n.d.). *Environmental economics.* https://www.epa.gov/environmental-economics

10. Sassen, S. (2001). The global city: New York, London, Tokyo. *Political Science Quarterly,* 107(2), 370–371. https://doi.org/10.2307/2152688

11. Banzhaf, A.S., & Walsh, R. P. (2008). Do people vote with their feet? An empirical test of Tierbout's mechanism. *American Economic Review,* 98(3), 843–863.

12. Wirth, L. (1938). Urbanism as a way of life. *The American Journal of Sociology,* 44, 1–24.

13. Lin, J., & Mele, C. (Eds). (2013). *The urban sociology reader.* Routledge.

14. Durkheim, E. (1893). *The division of labor in society.* The Free Press.

15. Lin & Mele (2013).

16. Jackson, K. (1985). *Crabgrass frontiers: The suburbanization of the United States.* Oxford University Press.

17. Smith, M. E. (2005). City size in late post classic Mesoamerica. *Journal of Urban History,* 31(4), 403–434.

18. Howard, E. (1938). *Garden cities of tomorrow.* Swann Sonnenschein & Co.

19. U.S. Census. (n.d.). *Annual Capital Expenditures Survey.* https://www.census.gov/programs-surveys/aces.html

20. Randolph, J. (2012). *Environmental Land Use Planning and Management: 2nd Edition.*. Island Press.

21. Intergovernmental Panel on Climate Change. (n.d.). *History of the IPCC.* https://www.ipcc.ch/about/history/

22. Ellis, J., & Alteen, D. (2021. January 21). *The Paris Climate agreements: What you need to know.* The New York Times. https://www.nytimes.com/2021/01/21/climate/biden-paris-climate-agreement.html

23. United Nations. (n.d.). Do you know all 17 SDGs? *History.* https://sdgs.un.org/goals

24. United Nations Developmental Programme. (2016, November 15). *From MDGs to sustainable development for all: Lessons from 15 years of practice.* https://www.undp.org/publications/mdgs-sustainable-development-all

25. United Nations. (2022). *The sustainable development goals report 2022.* https://unstats.un.org/sdgs/report/2022/

26. Jones-Roy, A., & Dottle, R. (2018, January 30). *We measured Trump's first year goals according to his own goals. Five Thirty Eight.* https://fivethirtyeight.com/features/we-measured-trumps-first-year-according-to-his-own-goals-heres-what-we-found/

27. OpenStax. (n.d.). *How economies can be organized. Principles of Economics.* https://openstax.org/books/principles-economics/pages/1-4-how-economies-can-be-organized-an-overview-of-economic-systems

28. Weatherford, J. (2016). *Genghis Khan and the quest of God: How the world's greatest conqueror gave us religious freedom.* Viking.

29. Cohan, P. (2021, October 26). *Why Tesla's market value will double to $2 trillion. Forbes.* https://www.forbes.com/sites/petercohan/2021/10/26/why-teslas-market-value-will-double-to-2-trillion/?sh=2d0ec3e47ccb

30. Environmental Protection Agency. (n.d.). *Reuse, reduce, recycle.* https://www.epa.gov/recycle

31. Menand, L. (2018, August 27). *Francis Fukuyama postpones the end of history.* The New Yorker. https://www.newyorker.com/magazine/2018/09/03/francis-fukuyama-postpones-the-end-of-history

Chapter 2: The Environmental and Health Impacts of Civilization

1. Khan, Z.B. (2015). Knowledge, human capital, and economic development: Evidence from the British industrial revolution, 1750–1930 (*NBER* Working Paper No. 20853). https://www.nber.org/papers/w20853

2. Environmental Protection Agency. (n.d.) *Frequently asked questions about climate change.* https://www.epa.gov/climatechange-science/frequently-asked-questions-about-climate-change#cold-snowfall

3. Shen, S. et al. (2011). Calibrating the end-Permian mass extinction. *Science, 334(6061),* 1367–1372.

4. Penn, J., Deutsch, C., Payne, J., & Sperling, E (2018). Temperature dependent hypoxia explains biogeography and severity of end-Permian marine mass extinction. *Science,* 362(6419).

5. Environmental Protection Agency. (2022). *Inventory of U.S. greenhouse gas emissions and sinks: 1990-2020: 1-841* https://www.epa.gov/system/files/documents/2022-04/us-ghg-inventory-2022-main-text.pdf

6. Intergovernmental Panel on Climate Change (IPCC). (2021). *Climate change 2021: The physical science basis. Sixth Assessment Report of the Intergovernmental Panel on Climate Change.* https://www.ipcc.ch/report/ar6/wg1/

7. NOAA/ESRL. (n.d.). *Trends in atmospheric carbon dioxide, methane, nitrous oxide.* https://gml.noaa.gov/

ccgg/trends/gr.html

8. EIA. (2022) *Monthly energy review, February 2022.* U.S. Department of Energy.

9. Cleveland Clinic. (2022 April 26) *How many calories should you eat in a day?* https://health.clevelandclinic.org/how-many-calories-a-day-should-i-eat/

10. Marion Nestle. (2002) *Food politics.* University of California Press.

11. U.S. Energy Information Administration. (2021). *Natural gas weekly update for week ending March 31, 2021.* https://www.eia.gov/naturalgas/weekly/archivenew_ngwu/2021/04_01/

12. Erisman, J. W., Sutton, M. A., Galloway, J., Klimont, Z., & Winiwarter, W. (2008). How a century of ammonia synthesis changed the world. *Nature Geoscience,* 1(10), 636–639.

13. NIEHS/NIH. (n.d.). *Algal blooms.* https://www.niehs.nih.gov/health/topics/agents/algal-blooms/index.cfm

14. WHO. (2021, June 9). *Malnutrition.* https://www.who.int/news-room/fact-sheets/detail/malnutrition

15. Pingali, P. (2012). Green revolution: Impacts, limits and the path ahead. *Proceedings of the National Academy of Sciences of the United States of America,* 109(31), 12302–12308.

16. FAO. (2021). *World food and agriculture—statistical yearbook 2021.* https://doi.org/10.4060/cb4477en

17. IPCC. (2022). *Summary for policymakers. In Climate change 2022: Impacts, adaptation and vulnerability: Contribution of Working Group II to the Sixth Assessment Report of the Intergovernmental Panel on Climate Change.* Cambridge University Press https://doi.org/10.1017/9781009325844.001

18. Environmental Protection Agency. (n.d.). *Global greenhouse emission data.* https://www.epa.gov/ghgemissions/global-greenhouse-gas-emissions-data

19. IPCC. (2014). *Climate change 2014: Mitigation of climate change.* Fifth Assessment Report of the Intergovernmental Panel on Climate Change. https://www.ipcc.ch/report/ar5/wg3/

20. Boden, T.A., Marland, G., & Andres, R. J. (2017). National CO2 emissions from fossil-fuel burning, cement manufacture, and gas flaring: 1751–2014. *Carbon Dioxide Information Analysis Center, Oak Ridge National Laboratory, U.S. Department of Energy.* https://doi.org/10.3334/CDIAC/00001_V2017

21. International Energy Agency. (2021). *Global energy review: CO2 emissions in 2021.* https://iea.blob.core.windows.net/assets/c3086240-732b-4f6a-89d7-db01be018f5e/GlobalEnergyReviewCO2Emissionsin2021.pdf

22. Gray, R. (2020). The extended Khardashev Scale. *The Astronomical Journal,* 159, 228.

23. U.S. Energy Information Administration. (2022). *Monthly energy review,* table 1.3. https://www.eia.gov/totalenergy/data/monthly/

24. U.S. Energy Information Administration. (n.d.). *Electricity explained: How electricity is generated.* https://www.eia.gov/energyexplained/electricity/how-electricity-is-generated.php

25. IEA. (2021). *Key world energy statistics 2021.* https://www.iea.org/reports/key-world-energy-statistics-2021/supply

26. WHO. (2019, November 12). *Safer water, better health.* https://www.who.int/publications-detail-redirect/9789241516891

27. UNICEF. (n.d.) *Water and the global climate crisis: 10 things you should know.* https://www.unicef.org/stories/water-and-climate-change-10-things-you-should-know

28. WHO. (2017 May 2). *Diarrhoeal disease.* https://www.who.int/news-room/fact-sheets/detail/diarrhoeal-disease

29. WHO & UNICEF. (2019 June) *Progress on household drinking water, sanitation and hygiene 2000–2017: Special focus on inequalities.* https://www.unicef.org/reports/progress-on-drinking-water-sanitation-and-hygiene-2019

30. Collier, S. A., Deng, L., Adam, E. A., Benedict, K. M., Beshearse, E. M., Blackstock, A. J., . . . &Beach, M. J. (2021). Estimate of burden and direct healthcare cost of infectious waterborne disease in the United States. *Emerging Infectious Diseases,* 27(1), 140–149. https://doi.org/10.3201/eid2701.190676.

31. Associated Press. (2022, September 27). *Current Mississippi governor previously held up funds for Jackson water crisis.* Eastbay Times. https://www.eastbaytimes.com/2022/09/27/current-governor-previously-held-up-funds-for-jackson-water-crisis/

32. Jones, B. (2022, September 1). *How Jackson, Mississippi ran out of water. Vox.* https://www.vox.com/2022/8/31/23329604/jackson-mississippi-water-crisis

33. Perry, A.M., Kane, J. W., & Romer, C. (2021, March 29) *In Jackson, Miss., a water crisis has revealed the racial costs of legacy infrastructure. Brookings.* https://www.brookings.edu/blog/the-avenue/2021/03/26/in-jackson-miss-a-water-crisis-has-revealed-the-racial-costs-of-legacy-infrastructure/

34. United Nations. (n.d.). *Sustainable Development Goal 6: Ensure availability and sustainable management of water and sanitation for all.* https://sdgs.un.org/goals/goal6

35. Dieter, C. A., Maupin, M. A., Caldwell, R. R., Harris, M. A., Ivahnenko, T. I., Lovelace, J. K., Barber, N. L., and Linsey, K. O. (2018). Estimated use of water in the United States in 2015. *U.S. Geological Survey Circular,* 1441. https://doi.org/10.3133/cir1441

36. World Health Organization. (2022, March 21). *Sanitation.* https://www.who.int/news-room/fact-sheets/detail/sanitation

37. Connor, R., & Koncagul, E. (2014). *The UN world water report 2014.* United Nations. https://digitallibrary.un.org/record/3892933

38. Randolph, J. (2012). Environmental land use: Planning and management. Island Press.

39. Mekonnen, M. M., & Hoekstra, A. Y. (2011). The green, blue and grey water footprint of crops and derived crop products, *Earth Syst. Sci,* 15, 1577–1600.

40. Water Footprint Network. (n.d.). *Product gallery.* https://waterfootprint.org/en/resources/interactive-tools/product-gallery/

41. *Gleick, P. H. (1993). Water in crisis: A guide to the world's fresh water resources. Oxford University Press.*

42. Searchinger, T., et al. (2019, July 19). *Creating a sustainable food future: A menu of solutions to feed nearly 10 billion people by 2050.* https://www.wri.org/research/creating-sustainable-food-future

43. Understanding Global Change. (n.d.). *Water cycle.* https://ugc.berkeley.edu/background-content/water-cycle/

44. The Economist (2019 March 2) *Thirsty Planet, 5-7.*

45. Food and Agricultural Organization. (n.d.). *World water resources by country* (table 3). https://www.fao.org/3/Y4473E/y4473e08.htm

46. United Nations. (2022, March 21), *The United Nation world water development report 2022: Groundwater making the invisible, visible.* https://www.unwater.org/publications/un-world-water-development-report-2022

47. UNESCO. (2021). The United Nation world water development report 2021: Valuing water. https://unesdoc.unesco.org/ark:/48223/pf0000375724

48. Organization for Economic Cooperation and Development. (n.d.). *Municipal waste, generation and treatment: Municipal waste generated per capita.* https://stats.oecd.org/Index.aspx?DataSetCode=MUNW

49. World Bank Group. (2018) *Municipal solid waste: A roadmap for reform for policy makers.* https://openknowledge.worldbank.org/handle/10986/30434?CID=SURR_TT_WBGCities_EN_EXT

50. Kaza, S., et al. (2018). *What a waste 2.0: A global snapshot of solid waste management to 2050.* https://openknowledge.worldbank.org/handle/10986/30317

51. Murad, M., & Chamhuria, S. (2007). Waste management and recycling practices of the urban poor: A case study in Kuala Lumpur. *Waste Management Research*, 25(1), 3–13.

52. EPA. (n.d.). *National overview: Facts and figures on materials, wastes and recycling.* https://www.epa.gov/facts-and-figures-about-materials-waste-and-recycling/national-overview-facts-and-figures-materials

53. Turrentine, J. (2019, July 12). *The US is the most wasteful country in the world.* https://www.nrdc.org/onearth/united-states-most-wasteful-country-world

54. Adejumo, I., & Adebiyi, O. (2020). *Agricultural solid wastes: Causes, effects, and effective management in strategies of sustainable solid waste management* (Ed. H. M. Saleh). https://www.intechopen.com/chapters/73517

55. UNEP. (2016, January 13). *Marine plastic debris & microplastics: Global lessons and research to inspire action and guide policy change.* https://www.unep.org/resources/publication/marine-plastic-debris-and-microplastics-global-lessons-and-research-inspire

56. Eriksen, M., et al. (2014). Plastic pollution in the world's oceans: More than 5 trillion plastic pieces weighing over 250,000 tons afloat at sea. *PLoS One*, 9, e111913.

57. Botterell, Z. L. R. (2018). Bioavailability and effects of microplastics on marine zooplankton: A review. *Environ. Pollut*, 245, 98–110.

58. Lim, X. Z. (2021, May 4). Microplastics are everywhere. *Nature*, 593(7857): 22-25. doi: 10.1038/d41586-021-01143-3. PMID: 33947993 https://www.nature.com/articles/d41586-021-01143-3#ref-CR2

59. NASA. (2004, June 6). *Ozone.* https://www.nasa.gov/audience/foreducators/postsecondary/features/F_Ozone.html

60. FAO, IFAD, UNICEF, WFP, & WHO (2022). *The state of food security and nutrition in the world 2022: Repurposing food and agricultural policies to make healthy diets more affordable.* https://doi.org/10.4060/cc0639en

61. CDC. (n.d.). *Adult obesity facts.* https://www.cdc.gov/obesity/data/adult.html

62. CAISO. (2023). How power flows in California. http://www.caiso.com/about/Pages/OurBusiness/How-power-flows-in-California.aspx

63. SCE. (2020). Reimagining the grid. https://download.newsroom.edison.com/create_memory_file/?f_id=5fcfb5f62cfac23b06eb7d39&content_verified=True

Chapter 3: History of Urbanization

1. Reich, S (1998). What is globalization: Four possible answers [Working paper 261]. *Kellogg Institute.*

2. Wilkinson, T.J., Philip, G.J., et al. (2007). Contextualizing early urbanization: Settlement cores, early states, and agro pastoral strategies in the fertile crescent during the 4th and 3rd millenia BC. J *World Prehis*, 27, 43–109

3. Massa, M., & Palmisano, A. (2018). Change and continuity in the long-distance exchange networks between western/central Anatolia, northern levant and northern Mesopotamia, c. 3200–1600 BCE. *J. Anthropol. Archaeol.*, 49, 65–87.

4. Benati, G., Guerriero, C., & Federico, Z. (2021). The economic and institutional determinants of trade expansión in Bronze Age greater Mesopotamia. *Journal of Archaeological Science*, 131, 105398

5. Kenoyer, J. (1997). Trade and technology from Indus Valley: New insights from Harrapa, Pakistan. *World Archaeology,* 29(2), 262–280.

6. Kenoyer, J. M., et al (2013). A new approach to tracking connection between the Indus Valley and Mesopotamia: Initial results of strontium analyses from Harrapa and Ur. *Journal of Archaeological Science,* 40, 2286–2297.

7. Malliya, P., et al. (2014). Tamil merchants in ancient Mesopotamia. *PLOS One,* 9(10), e109331.

8. Lawler A (2008 Jun 6) Indus Collapse: The End or the Beginning of an Asian Culture?. Science, 320(5881),1281-1283.

9. Wallerstein, I. (1974). *The modern world system I: Capitalist agriculture and the origins of the European world-economy in the sixteenth century.* Academic Press.

10. Cobb, M. (2015). The chronology of Roman trade in the Indian Ocean from Augustus to early 3rd century CE. *Journal of the Economic and Social History of the Orien*t, 58, 362–418.

11. Goldfrank, W. L. (2000). Paradigm regained? The rules of Wallerstein's world system method. *Journal of World-Systems Research*, 6(2), 150–195.

12. Pitts, M., & Versluys, M. J. (2015). *Globalisation and the Roman world: World history, connectivity and material culture.* Cambridge University Press.

13. Blakemore, E (2019, June). Who were the Mongols? National Geographic. https://www.nationalgeographic.com/culture/article/mongols

14. Knutson, S. A. (2020). Archaeology and the silk road model. *World Archaeology,* 52(4), 619–638.

15. Hedenstierna-Jonsson, C., & Holmquist Olausson, L. (2006). *The Oriental mounts from Birka's Garrison: An expression of warrior rank and status.* Publisher Unknown.

16. Whitfield, S. (Ed.). (2019). *Silk roads: Peoples, cultures, landscapes.* University of California Press.

17. Pyne, S. (2021). *The great ages of discovery: How Western civilization learned about a wider world.* University of Arizona Press

18. Wesseling, H. (2009). Globalization: A historical perspective. *European Review,* 17(3), 455–462.

19. Janvry, A. (2021). *Development economics: Theory and practice.* Taylor & Francis.

20. Goldstone, J. (1996). Gender, work and culture: Why the industrial revolution came early to England but late to China. *Sociological Perspectives*, 39(1), 1–21.

21. Dung, N., & Signe, L. (2020). *The 4th industrial revolution and Africa.* Brooking Institution.

22. Nash, T. (2021). *Timeless values: The British industrial revolution.* Northwood University. https://www.northwood.edu/afeu/when-free-to-choose/timeless-values-the-british-industrial-revolution-1750-1830

23. Hoppit, J. (1987). Understanding the industrial revolution. *The Historical Journal*, 1, 211–224.

24. White, M. (2009). The industrial revolution. *The British Library*. https://www.bl.uk/georgian-britain/articles/the-industrial-revolution

25. Min, J., et al. (2019). The 4th industrial revolution and its impact on occupational health and safety:

Worker's compensation and labor conditions. *Saf Health Work,* 10(4), 400–408.

26. Frey, C. B., & Osborne, M. A. (2017). The future of employment: How susceptible are jobs to computerisation? *Technol Forecast Soc Chang,* 114, 254–280. http://www.sciencedirect.com/science/article/pii/S0040162516302244

27. Crafts, N. (2022). Slow real-wage growth during the industrial revolution: Productivity paradox or pro-rich growth. *Oxford Economic Papers,* 74(1), 1–13.

28. Engineering and Technology History Wiki. (n.d.). Telegraph. https://ethw.org/Telegraph

29. Snowdon, B. (2004). Globalization in historical perspective [Book review]. *Southern Economic Journal,* 71(1), 201–204.

30. Strikweda, C. (2016). World War I in the history of globalization. *Historical Reflections,* 42(3), 112–132

31. The Economist. (2017, June 14). The globalization counter reaction. The Economist, 24-26.

32. Tassava, C. J. (n.d.). The American economy during WWII. *Economic History Association.* https://eh.net/encyclopedia/the-american-economy-during-world-war-ii/

33. Mayer, B. (2009). Cross movement coalition formation: Bridging the labor-environment divide. *Sociological Inquiry,* 79(2), 219–239.

34. NPR. (2022 April 22). Oil traders help Russia still make billions from crude exports https://www.npr.org/2022/04/22/1092591767/despite-u-s-sanctions-oil-traders-help-russian-oil-reach-global-markets

35. Masterson, V. (2022 April). Russian and Ukrainian exports to the world. World Economic Forum. https://www.weforum.org/agenda/2022/04/world-bank-ukraine-food-energy-crisis/

Chapter 4: Economic Fundamentals

1. Gibson-Light, M. (2018). Ramen politics: Information inmate economy in the contemporary American prison. *Qualitative Sociology,* 41, 199–220.

2. Taleb, N. (2020). Bitcoin, currencies and eragility. *Quantitative Finance, 1-6. https://doi.org/10.48550/arXiv.2106.14204*

3. Pikkety, T. (2014). *Capital in the 21st century.* Belknap Press.

4. Overseen, E. (2021). Capitalism and alienation: Towards a Marxist theory of alienation for the 21st century. *European Journal of Social Theory,* 25(3): 1–18

5. Bricker, J., Henriques, A., Krimmel, J., & Sabelhaus, J. (2016). Measuring income and wealth at the top using administrative and survey data. *Brookings Papers on Economic Activity,* 1, 261–321.

6. Feiveson, L., & Sabelhaus, J. (2018). How does intergenerational wealth transmission affect wealth concentration? *Board of Governors of the Federal Reserve System.* https://www.federalreserve.gov/econres/notes/feds-notes/how-does-intergenerational-wealth-transmission-affect-wealth-concentration-20180601.htm

7. Rotta, T.M. (2022). Information rents, economic growth and inequality: An empirical study of the US. *Cambridge Journal of Economics,* 46(2), 341–370.

8. Ryan-Collins, J., Mazzuvato, M., & Gouzoulis, G. (2020). Theorizing and mapping modern economic rents [Working Paper 2020-13]. *UCLA Institute for Innovation and Public Use.*

9. Murray, C. (2018). The state of US health, 1990–2016: Burden of disease, injuries, and risk factors among US states. *JAMA,* 319(14), 1444-1472.

10. American Psychological Association. (2015) *Stress in America: Paying with our health.* https://www.apa.org/news/press/releases/stress/2014/stress-report.pdf

11. Burkhaser, R., DeNeve, J. E., & Powdthavee, N. (2015). Top incomes and human well-being around the world [*IZA discussion paper No. 9677*]. https://papers.ssrn.com/sol3/papers.cfm?abstract_id=2725038

12. Dabla-Norris, E., et al. (2015). Causes and consequences of global inequality. *International Monetary Fund.* https://www.imf.org/external/pubs/ft/sdn/2015/sdn1513.pdf

13. McGee, R. (1989). The economic thought of David Hume. *Hume Studies,* 15(1), 184–204.

14. Board of Governors of the Federal Reserve System. (n.d.). *Monetary policy: What are its goals? How does it work?* https://www.federalreserve.gov/monetarypolicy/monetary-policy-what-are-its-goals-how-does-it-work.htm

15. Board of Governors of the Federal Reserve System. (n.d.). *What is the money supply? Is it important?* https://www.federalreserve.gov/faqs/money_12845.htm

16. Hall, P., & Soskice, D. (2001). Varieties of capitalism: Institutional foundations of comparative advantage. *The Academy of Management Review,* 28(3). https://scholar.harvard.edu/hall/publications/varieties-capitalism-institutional-foundations-comparative-advantage

17. Hope, D., & Limberg, J. (2022). The economic consequences of major tax cuts for the rich, Socio-Economic Review, 20(2), 539–559. https://doi.org/10.1093/ser/mwab061

18. Guttman, A. (2022, February 25). US advertising industry: Statistics and facts. *Statista.* https://www.statista.com/topics/979/advertising-in-the-us/

19. Hamadeh, N., Van Rompaey, C., & Metreau, E. (2021). New World Bank country classifications by income level: 2021-2022. *Data Blog.* https://blogs.worldbank.org/opendata/new-world-bank-country-classifications-income-level-2021-2022

Chapter 5: The Institutions of Finance and Trade

1. Heinzel, M., et al. (2021). Birds of a feather? The determinants of impartiality perceptions of the IMF and the World Bank. *Review of International Political Economy,* 28(5), 1249–1273.

2. Rapley, J. (2007). *Understanding development, theory and practice in the 3rd world.* Lynne Reinner.

3. Duignan, P., & Gann, L. H. (1997). *The Marshall plan.* Hoover Digest. https://www.hoover.org/research/marshall-plan

4. International Monetary Fund. (n.d.). *What is the IMF?* https://www.imf.org/en/About/Factsheets/IMF-at-a-Glance

5. Federal Reserve History. (n.d.). *Creation of the Bretton Woods system.* https://www.federalreservehistory.org/essays/bretton-woods-created

6. The Federal Reserve Bank of Dallas. (2011) *Globalization and Monetary Policy Institute annual report 2011.* https://fraser.stlouisfed.org/title/annual-report-federal-reserve-bank-dallas-475/2011-annual-report-596535

7. IMF. (2009). *Balance of payments and international, investment position manual* (6th ed). https://www.imf.org/external/pubs/ft/bop/2007/pdf/bpm6.pdf

8. Bureau of Economic Analysis (n.d.). Home page. www.bea.gov

9. Stiglitz, J. (2003). Globalization and its discontents. Norton.

10. IMF. (2021 April 29). *At a glance: The IMF's firepower.* https://www.imf.org/en/About/infographics/imf-firepower-lending

11. IMF. (2022). *Active lending commitments as of May 2022.* https://www.imf.org/external/np/fin/tad/extarr11.aspx?memberKey1=ZZZZ&date1key=2025-12-31

12. World Bank. (n.d.). *World Bank: Project and operations.* https://projects.worldbank.org/en/projects-operations/projects-home

13. World Bank. (n.d.). *World Bank: Projects and operations map.* https://maps.worldbank.org/

14. The World Bank. (2021). *Annual report: From crisis to green, resilient and inclusive recovery.* https://documents.worldbank.org/en/publication/documents-reports/documentdetail/120541633011500775/the-world-bank-annual-report-2021-from-crisis-to-green-resilient-and-inclusive-recovery

15. World Trade Organization. (2021). *World trade report 2021: Economic resilience and trade.* https://www.wto.org/english/res_e/booksp_e/wtr21_e/00_wtr21_e.pdf

16. The World Trade Organization. *World Trade Report (2021): Economic Resilience and Trade.* https://www.wto.org/english/res_e/booksp_e/wtr21_e/00_wtr21_e.pdf

17. World Trade Organization. (n.d.). *Stats.* https://Stats.wto.org

18. Bureau of Economic Affairs. *US trade data.* https://www.bea.gov/data/intl-trade-investment/international-trade-goods-and-services

19. OECD. (n.d.). *Discover the OEC: Better policies for better lives.* https://www.oecd.org/general/Key-information-about-the-OECD.pdf

20. Klein, L., & Salvatore, D. (2013). Shift in the world economic center of gravity from G7 to G20. *Journal of Policy Modelling, 35,* 416–424.

21. Soberl, M. (2021 June 10). The G7 finds new purpose as G20 takes a back seat. *Official Monetary and Financial Institution Forum.* https://www.omfif.org/2021/06/g7-finds-new-purpose-as-g20-takes-back-seat/

22. Henderson, D. (n.d.). David Ricardo. *Econlib.* https://www.econlib.org/library/Enc/bios/Ricardo.html

23. Pettinger, T. (2019, June 28). Economic efficiency. Economics: Helping to simplify economics https://www.economicshelp.org/microessays/costs/efficiency/

24. Boughton, J. M. (2021). *Harry White and the American creed: How a federal bureaucrat created the modern global economy (and failed to get the credit).* Yale University Press.

25. Federal Reserve. (n.d.). *Stress testing under the prior capital frameworks.* https://www.federalreserve.gov/supervisionreg/stress-tests-capital-planning.htm

26. Federal Reserve. (n.d.). *Policy tools: Reserve requirements.* https://www.federalreserve.gov/monetarypolicy/reservereq.htm

27. Carliner, M. S. (1998). Development of federal homeownership policy. *Housing Policy Debate, 9*(2), 299–321.

28. Yescombe, E. R. (2014). *Principles of project finance* (2nd ed.). Academic Press & Elsevier.

29. SIFMA. (n.d.). *Research quarterly: Fixed income—issuance and trading, first quarter 2021.* https://web.archive.org/web/20210513170717/https://www.sifma.org/resources/research/research-quarterly-fixed-income-issuance-and-trading-first-quarter-2021/

30. U.S. Treasury. (n.d.). *Major foreign holders of Treasury securities* https://ticdata.treasury.gov/resource-

center/data-chart-center/tic/Documents/mfh.txt

31. Dugga, W. (2022, April 18). *Securities defined.* https://money.usnews.com/investing/term/securities

32. Hicks, C. (2020, March 23). *What are derivatives and should you invest in them. U.S. News.* https://money.usnews.com/investing/investing-101/articles/what-are-derivatives-and-should-you-invest-in-them

33. Business Wire. (2021 March 10). *Global financial services and market outlook (2021–2030).* https://www.businesswire.com/news/home/20210310005386/en/Global-Financial-Services-Market-Outlook-2021-2030-Expected-to-Reach-28.52-Trillion-by-2025—ResearchAndMarkets.com

34. Pikkety, T. (2014). *Capital in the 21st century.* Belknap Press.

35. IMF. (n.d.). *Where the IMF gets its money.* https://www.imf.org/en/About/Factsheets/Where-the-IMF-Gets-Its-Money

Chapter 6: Demography

1. Smith, S., Tayman, J., & Swanson, D. (2001). State and local population projections, methodology and analysis. Kluwer Academic.

2. Hackett, C. (2022, July 21). *Global population projected to exceed 8 billion in 2022; half live in just seven countries. Pew Research Center.* https://www.pewresearch.org/fact-tank/2022/07/21/global-population-projected-to-exceed-8-billion-in-2022-half-live-in-just-seven-countries/

3. UN Department of Economic and Social Affairs, Population Division. (2022). *World population prospects 2022: Summary of results.*

4. Reuters. (2022 June 3). *Japan recorded record low births, biggest ever population drop in 2021* https://www.reuters.com/world/asia-pacific/japan-recorded-record-low-births-biggest-ever-population-drop-2021-2022-06-03/

5. U.S. Census. (2020). *Demographic turning points for the US: Population projections for 2030 to 3060.* https://www.census.gov/library/publications/2020/demo/p25-1144.html

6. Population Reference Bureau. (n.d.). *World population data sheet 2021.* https://interactives.prb.org/2021-wpds/

7. Dunne, T. (2012, June 23). *Household formation and the great recession. Federal Reserve Bank of Cleveland.* https://www.clevelandfed.org/en/newsroom-and-events/publications/economic-commentary/economic-commentary-archives/2012-economic-commentaries/ec-201212-household-formation-and-the-great-recession.aspx

8. Garcia, D., & Paciorek, A. (2020, August 7). *An early evaluation of the effects of the pandemic on living arrangement and household formation.* Board of Governors of the Federal Reserve. https://www.federalreserve.gov/econres/notes/feds-notes/an-early-evaluation-of-the-effects-of-the-pandemic-on-living-arrangements-and-household-formation-20200807.html

9. Skakkebaek, N., et al. (2006). Is human fecundity declining? Int J Androl, (1), 2–11.

10. History.com Editors. (2020, January 30). *Pandemics that changed history.* https://www.history.com/topics/middle-ages/pandemics-timeline

11. Thornton, R. (1997). Aboriginal North American population and rates of decline, ca. A.D. 1500–1901. *Curr. Anthropol.,* 38, 310–315.

12. Munnell, A. H. (2021 September). *Social Security's financial outlook: The 2021 update in perspective.* https://crr.bc.edu/briefs/social-securitys-financial-outlook-the-2021-update-in-perspective/

13. Population Reference Bureau. (2012, November 7). *South Korea's demographic dividend.* *https://www.prb.org/resources/south-koreas-demographic-dividend/*

14. Kearney, M.S., & Levine, P. (2021, May 24). *Will births in the US rebound? Probably not. Brookings.* https://www.brookings.edu/blog/up-front/2021/05/24/will-births-in-the-us-rebound-probably-not/

15. Hamilton, B. E., et al. (2021). Births: Provision data for 2020. *NVSS, Vital Statistics Rapid Release,* 12, 1–11.

16. CDC. (n.d.). *National Center for Health Statistics.* https://www.cdc.gov/nchs/index.htm

17. Loughran, D., & Zissimopolos, J. (2004). *Are there gains to delaying marriage? The effect of age at first marriage on career development and wages (Working paper series, WR-207).* RAND.

18. Skakkebaek, N. E., et al. (2021). Environmental factors in declining human fertility. *Nature Reviews Endocrinology,* 18, 139–157.

19. Martine, G., McGranahan, G., Montgomery, M., & Fernandez-Castilla, R. (2008). *The new global frontier: Urbanization, poverty and environment in the 21st century.* 2008. Routledge.

20. United Nations, Department of Economic and Social Affairs, Population Division. (2019). *World urbanization prospects: The 2018 revision* (ST/ESA/SER.A/420). https://www.un.org/development/desa/pd/news/world-population-prospects-2019-0

21. UN Habitat. (2021, July 19). *Proportion of urban population living in slum households by country or area 1990–2018.* https://data.unhabitat.org/datasets/proportion-of-urban-population-living-in-slum-households-by-country-or-area-1990-2018-percent/explore

22. UN Habitat. (2003). *Slums of the world: The face of urban poverty in the new millennium.* https://unhabitat.org/slums-of-the-world-the-face-of-urban-poverty-in-the-new-millennium

23. Grubler, A., & Fisk, D. (2013*). Energizing sustainable cities: Assessing Urban Energy.* Routledge.

24. Fink, G., Günther, I., & Hill, K. (2014). Slum residence and child health in developing countries. *Demography,* 51, 1175–1197.

25. U.S. Census Bureau .(2021 August 12). 2020 Census frequently asked questions about race and ethnicity. https://www.census.gov/programs-surveys/decennial-census/decade/2020/planning-management/release/faqs-race-ethnicity.html

26. National Vital Statistics Report (2021). *United States life tables, 2018. US Department of Health and Human Services,* 70(1), 1–18.

27. Ibid.

28. U.S. Department of Health and Human Services. (n.d.). *Instructions for completing the cause of death section.* https://www.cdc.gov/nchs/data/dvs/blue_form.pdf

29. Congressional Research Service. (2016, January 7). *China's Hokou system: Reform and economic implications.* *https://crsreports.congress.gov/product/pdf/IF/IF10344/3*

30. United Nations. (2007). *Indicators of sustainable development: Guidelines and methodologies (3rd ed.).* https://sustainabledevelopment.un.org/index.php?page=view&type=400&nr=107&

31. The Economist. (2023, July). The new Asian family (2023 Aug 2) https://china.usc.edu/sites/default/files/forum/economist-2023-07-the-new-asian-family-rejecting-confucian-family.pdf

32. Mellor, S. (2022, December 13). Gen Z and young Millennials have found a new way to afford luxury

handbags—living with mom and dad. Fortune. https://fortune.com/2022/12/13/gen-z-young-millennials-way-afford-luxury-handbags-watches-living-mom-dad-morgan-stanley/

33. WHO. (n.d.). Coronavirus dashboard. https://covid19.who.in

34. Beine, M., Docquier, F., & Rapoport, H. (2001). Brain drain and economic growth: Theory and evidence. Journal of Development Economics, 64(1), 275–289.

Chapter 7: Poverty

1. Moffit, R. A., & Zahn, M. V. (2019). The marginal labor supply disincentives of welfare: Evidence from administrative barriers to participation (*Working paper 26028*). *National Bureau of Economic Research*. *https://doi.org/10.3386/w26028*

2. Loprest, P., Schmidt, S., & Witte, A. (2000). Welfare reform under PRWORA: Aid to children with working families. *NBER: Tax Policy and the Economy, 14, 157–203.*

3. Tanner, M., & Highes, M. (2015). Policy Brief: The work vs welfare trade off: Europe. *CATO Institute Policy Analysis,* (779). https://clips.cato.org/sites/default/files/iea%20welfare.pdf

4. Ingram, J., & Dublois, H. (2021, November 11). *How the era of expanded welfare programs is keeping Americans from working.* FGA. https://thefga.org/paper/expanded-welfare-keeping-americans-from-working/

5. Urbanik, J., & Chensvold, L. (2021, August 30). Workers are at the heart of our economic recovery. *National Fund for Workforce Solutions.* https://nationalfund.org/help-workers-succeed/?gclid=Cj0KCQjwy5maBhDdARIsAMxrkw2MAMBW5QkulXAjkiT_ZWa8UWsj6B2yQSAG7E Y8XPoQ98XAlr65edEaAiOVEALw_wcB

6. Edelman, P. (2012). *So rich, So poor. Why the world's wealthiest nation in the world is losing the battle against poverty.* The New Press.

7. Dabla-Norris, E., et al. (2015). *Causes and consequences of global inequality.* International Monetary Fund. https://www.imf.org/external/pubs/ft/sdn/2015/sdn1513.pdf

8. Danziger, S., & Haveman, R. (2001). *Understanding poverty.* Russell Sage Foundation.

9. Gennetian, L. A., Hirsh-Pasek, K., & Yoshikawa, H. (2020, July 15). *Progress on social mobility takes more than two viewpoints. Brookings.* https://www.brookings.edu/blog/education-plus-development/2020/07/ 15/progress-on-social-mobility-takes-more-than-two-viewpoints/

10. Lecerf, M. (2016). *Poverty in the European Union.* European Parliamentary Research Service. https://epthinktank.eu/2016/03/22/poverty-in-the-european-union-the-crisis-and-its-aftermath/

11. Shrider, A. S., et al. (2021). *Current population reports P60-273, Income and poverty in the US 2020.* U.S. Census Bureau. https://www.census.gov/library/publications/2021/demo/p60-273.html

12. Fisher, G. (1992). The development and history of the poverty thresholds. *Social Security Bulletin,* 55(4), 3–14.

13. U.S. Census Bureau. (2020). *Decennial and ACS 5-year estimates.* https://www.census.gov/data/developers/ data-sets/acs-5year.html

14. U.S. Census Bureau. (n.d.). Current population survey. https://www.census.gov/programs-surveys/ cps.html

15. OECD. (2021). *Is the German middle class crumbling? Risks and opportunities.* https://www.oecd.org/

employment/German-Middle-Class-2021-Highlights_web.pdf

16. Hamadeh, N., Van Rompaey, C., & Metreau, E. (2021). *New World Bank country classifications by income level: 2021–2022.* Data Blog. https://blogs.worldbank.org/opendata/new-world-bank-country-classifications-income-level-2021-2022

17. United Nations. (2020, July 7). *Covid-19 is reversing decades of progress on poverty, healthcare and education.* https://www.un.org/development/desa/en/news/sustainable/sustainable-development-goals-report-2020.html

18. Kluegel, J. R., & Smith, E. R. (1986). Beliefs about inequality: American's view of what is and what ought to be. Aldine de Gruyter.

19. Cozzarelli, C., Wilkinson, A. V., & Tagler, M. J. (2001). Attitudes towards the poor and attributions for poverty. *Journal of Social Issues, 57*(2), 207–227.

20. Schneider, H. L., & Ingram, H. (1997). *Policy design for democracy.* University Press of Kansas.

21. Hunt, M. O. (1996). The individual, society or both? A comparison of Black, Latino and White beliefs about the causes of poverty. *Social Forces, 75*(1), 293–322.

22. United Nations. (n.d.). *Least developed countries.* https://www.un.org/ohrlls/content/least-developed-countries

23. World Bank. (2022, September 14). *Measuring poverty* https://www.worldbank.org/en/topic/measuringpoverty#2

24. World Bank. (2015, September 30). *Global poverty line update* https://www.worldbank.org/en/topic/poverty/brief/global-poverty-line-faq

25. U.S. Department of Housing and Urban Development. (n.d.). *Overview of HUD Section 8 limits.* https://www.huduser.gov/portal/datasets/il//il22/IncomeLimitsMethodology-FY22.pdf

26. U.S. Department of Housing and Urban Development. (n.d.). *Income limits.* https://www.huduser.gov/portal/datasets/il.html#2022_query

27. World Bank. (2020). *Poverty and shared prosperity 2020: Reversals of fortune.* https://doi.org/10.1596/978-1-4648-1602-4

28. World Bank Group, Development Research Center of the State Council, & the People's Republic of China. (2022). *Four decades of poverty reduction in China.* https://thedocs.worldbank.org/en/doc/bdadc16a4f5c1c88a839c0f905cde802-0070012022/original/Poverty-Synthesis-Report-final.pdf

29. Yueh, L. (2015, December 9). China's growth: A brief history. Harvard Business Review. https://hbr.org/2015/12/chinas-growth-a-brief-history

30. Rapley, J. (2007). *Understanding development, theory and practice in the 3rd world.* Lynne Reinner.

31. Isbister, J. (2006). *Promises not kept.* Kumarian Press.

32. Lin, J. (2015). The Washington consensus revisited: A new structural economics perspective. *Journal of Economic Policy Reform, 18*(2), 96–113.

33. Easterly, W. (2001). The lost decades: Stagnation in spite of policy reform. *Journal of Economic Growth, 6,* 135–157.

34. Archibong, B., Coulibaly, B. S., & Okonjo-Iweala, N. (2021, February 19). *How have the Washington consensus reforms affected economic performance in sub-Saharan Africa?* Brookings. https://www.brookings.edu/blog/africa-in-focus/2021/02/19/how-have-the-washington-consensus-reforms-affected-economic-performance-in-sub-saharan-africa/

35. Congressional Research Services. (2022, March 24) *The TANF block grant.* https://crsreports.congress.gov/product/pdf/IF/IF10036

36. U.S. Census Bureau, Current Population Reports. (n.d.). *The supplemental poverty measure:2018 (P60-268).*

37. USDHHS. (2021, June 5) *New HHS data show more Americans than ever have health ccoverage through the ACA.* https://www.census.gov/library/publications/2019/demo/p60-268.html

38. CMS. (2021, December 21). *CMS releases late enrollment figures for Medicare, Medicaid and CHIP.* https://www.cms.gov/newsroom/news-alert/cms-releases-latest-enrollment-figures-medicare-medicaid-and-childrens-health-insurance-program-chip

39. CDC & NCHS. (n.d.). *Health expenditures.* https://www.cdc.gov/nchs/fastats/health-expenditures.htm

40. Yonzan, N., et al. (2022, January 18). *The impact of COVID-19 on poverty and inequality: Evidence from phone surveys.* Data Blog. https://blogs.worldbank.org/opendata/impact-covid-19-poverty-and-inequality-evidence-phone-surveys

41. Ramos, F. M., & Lara, J. (2022, March 18). *COVID-19 and poverty vulnerability.* Brookings. https://www.brookings.edu/blog/future-development/2022/05/18/covid-19-and-poverty-vulnerability/

42. Lara, J., & Mendez-Ramos, F. (2021). Poverty vulnerability: The role of poverty lines in the post pandemic era. *Economics Bulletin*, 41(4), 2690–2696. http://www.accessecon.com/Pubs/EB/2021/Volume41/EB-21-V41-I4-P232.pdf

43. World Bank. (2022) *Global economic prospects, June 2022.* https://doi.org/10.1596/978-1-4648-1843-1

44. Jolliffe, D. M., et al. (2022). *The impact of the 2017 PPPs on the international poverty line and global poverty* (Policy research working paper 9941). World Bank Group. http://documents.worldbank.org/curated/en/353811645450974574/Assessing-the-Impact-of-the-2017-PPPs-on-the-International-Poverty-Line-and-Global-Poverty

45. Department of Economic and Social Affairs, Statistics Division. (2021). The Sustainable Development Goals report: 2021. *United Nations.* https://unstats.un.org/sdgs/report/2021/goal-01/

46. The Economist (2020 May 16). *Is Zipcode Destiny, 23-24.*

47. Allas, T., Maksumainen, J., Manyika, J., & Navjot, S. (2020, September 15). *An experiment to inform universal basic income.* McKinsey & Company. https://www.mckinsey.com/industries/public-and-social-sector/our-insights/an-experiment-to-inform-universal-basic-income

Chapter 8: Public Health

1. Hanckock, B. H. (2018). Michael Foucault and the problematics of power: Theorizing DTCA and medicalized subjectivity. *Journal of Medicine and Philosophy: A Forum for Bioethics and Philosophy of Medicine, 43(4), 439–468.*

2. Center for Medicaid and Medicare Services. (n.d.). *National health expenditures 2021 highlights.* https://www.cms.gov/files/document/highlights.pdf

3. Word Health Organization. (2021, December 15). *Global expenditure on health: Public spending on the rise?* https://www.who.int/publications/i/item/9789240041219

4. NIH. (2016, June 22). *Americans spent $30.2 billion out-of-pocket on complementary health approaches.* https://www.nccih.nih.gov/news/press-releases/americans-spent-302-billion-outofpocket-on-

complementary-health-approaches

5. United Nations. (n.d.). *End hunger, achieve food security and improved nutrition and promote sustainable agriculture.* https://sdgs.un.org/goals/goal2

6. WHO. (2021, June 9). *Malnutrition.* https://www.who.int/news-room/fact-sheets/detail/malnutrition

7. WHO. (2022, February 3). *21st century health challenges: Can the essential public health functions make a difference?* https://www.who.int/publications/i/item/9789240038929

8. Public Health National Center for Innovations. (2020). *The 10 essential public health services.* https://phnci.org/uploads/resource-files/EPHS-English.pdf

9. Dicker, R., et al. (2006). *Principles of epidemiology (3rd ed.).* CDC.

10. NACCHO. (2019). *National profile of local health departments.* https://www.naccho.org/uploads/downloadable-resources/Programs/Public-Health-Infrastructure/NACCHO_2019_Profile_final.pdf

11. CDC. (n.d.). *What is epidemiology?* https://www.cdc.gov/training/publichealth101/epidemiology.html

12. Centers for Disease Control and Prevention. (n.d.). *Picture of America.* https://www.cdc.gov/pictureofamerica/pdfs/picture_of_america_prevention.pdf

13. U.S. Department of Health and Human Services, Office of Disease Prevention and Health Promotion. (n.d.). *Social determinants of health.* https://health.gov/healthypeople/objectives-and-data/social-determinants-health

14. Hahn, R. (1995). *Sickness and healing: An anthropological perspective.* Yale University Press.

15. The Hastings Center. (2019, January 4). *What does dead mean? How should we define death?* https://neurosciencenews.com/death-redefined-10426/

16. Centers for Disease Control and Prevention. (n.d.). *Facts about hypertension.* https://www.cdc.gov/bloodpressure/facts.htm#:~:text=Nearly%20half%20of%20adults%20in,are%20taking%20medication%20for%20hypertension

17. Mayo Clinic. (n.d.). *Fibromyalgia.* https://www.mayoclinic.org/diseases-conditions/fibromyalgia/symptoms-causes/syc-20354780

18. Hans, B., Merril, S., & Ida, S. (1997). *Medical anthropology and the world system: A critical perspective.* Bergin and Garvey.

19. CDC. (n.d.). *Measures of risk.* https://archive.cdc.gov/www_cdc_gov/csels/dsepd/ss1978/lesson3/section6.html

20. World Health Organization. (n.d.). *Constitution of the World Health Organization.* https://apps.who.int/gb/bd/PDF/bd47/EN/constitution-en.pdf?ua=1

21. World Health Organization. (n.d.). *Health and well-being.* https://www.who.int/data/gho/data/major-themes/health-and-well-being

22. WHO. (2022, June 16). *World mental health report: Transforming mental health for all.* https://www.who.int/publications/i/item/9789240049338

23. WHO. (n.d.). *Suicide rates.* https://www.who.int/data/gho/data/themes/mental-health/suicide-rates

24. Ehlman, D. C., Yard, E., Stone, D. M., Jones, C. M., & Mack, K. A. (2022). Changes in suicide rates—United States, 2019 and 2020. *Morb Mortal Wkly Rep*, 71, 306–312. http://dx.doi.org/10.15585/mmwr.mm7108a5

25. Schuch, F. B., & Vancampfort, D. (2021). Physical activity, exercise, and mental disorders: It is time to

move on. *Trends Psychiatry Psychother, 43*(3), 177–184.

26. Hottin, K., & Roder, B. (2013). Beneficial effects of physical exercise on neuroplasticity and cognition. *Neurisco Biobehav Rev, 37*, 2243–2257.

27. Feter, N., et al. (2022). Physical activity and the incidence of depression. *Mental Health Phys Act, 23*, 100468. https://doi.org/10.1016/j.mhpa.2022.100468

28. Codella, R., Terruzzi, I., & Luzi, L. (2016). Sugars, exercise and health. *J Affect Disord, 224*, 76–86. https://doi.org/10.1016/j.jad.2016.10.035

29. U.S. Department of Health and Human Services, Office of Disease Prevention and Health Promotion. (n.d.). *Social determinants of health.* https://health.gov/healthypeople/objectives-and-data/social-determinants-health

30. O'Neil, B. A., Forsythe, M. E., & Stanish, W. D. (2001). Chronic occupational repetitive strain injury. *Canadian Family Physician, 47*(2), 311–316.

31. U.S. Department of Labor, Bureau of Labor Statistics. (2019). *Table A-7: Fatal occupational injuries by worker characteristics and event or exposure, all United States.* https://www.bls.gov/iif/oshwc/cfoi/cftb0333.htmimage

32. Dooley, D., Fielding, J., & Levi, L. (1996). Health and unemployment. *Annual Review of Public Health, 17*, 449–465.

33. Freudenberg, N., & Ruglis, J. (2007). Reframing school dropout as a public health issue. *Preventing Chronic Disease, 4*(4), A107.

34. Krueger, P. M., Tran, M. K., Hummer, R. A., & Chang, V. W. (2015). Mortality attributable to low levels of education in the United States. *PloS One, 10*(7), e0131809.

35. Ross, C. E., & Wu, C. L. (1995). The links between education and health. *American Sociological Review, 60*(5), 719–745.

36. Baicker, K., Taubman, S. L., Allen, H. L., Bernstein, M., Gruber, J. H., Newhouse, J. P., Schneider, E. C., Wright, B. J., Zaslavsky, A. M., & Finkelstein, A. N. (2013). The Oregon experiment—Effects of Medicaid on clinical outcomes. *New England Journal of Medicine, 368*(18), 1713–1722.

37. Kutner, M., Greenburg, E., Jin, Y., & Paulsen, C. (2006). *The health literacy of America's adults: Results from the 2003 National Assessment of Adult Literacy (NCES 2006-483). Institute of Education Sciences,* National Center for Education Statistics. https://eric.ed.gov/?id=ED493284

38. California Center for Public Health Advocacy, PolicyLink, & UCLA Center for Health Policy Research. (2008, April). *Designed for disease: The link between local food environments and obesity and diabetes.* https://healthpolicy.ucla.edu/publications/Documents/PDF/Designed%20for%20Disease%20The%20Link%20Between%20Local%20Food%20Environments%20and%20Obesity%20and%20Diabetes.pdf

39. Ahern, M., Brown, C., & Dukas, S. (2011). A national study of the association between food environments and county-level health outcomes. *Journal of Rural Health, 27*(4), 367–379.

40. EPA. (n.d.). Learn about environmental justice. https://www.epa.gov/environmentaljustice/learn-about-environmental-justice

41. Gray, W., Shad, R. J., & Wolverton, A. (2010). *Environmental justice: Do poor and minority populations face more hazards? (Working paper series #10-10).* EPA. https://www.epa.gov/environmental-economics/working-paper-environmental-justice-do-poor-and-minority-populations-face

42. Baden, B., Noonan, D., & Turaga, R. (2007). Scales of justice: Is there a geographic bias in environmental equity analysis? *Journal of Environmental Planning and Management, 50,* 163–185.

43. Boer, T., Pastor, M., Sadd, J., & Snyder, L. (1997). Is there environmental racism? The demographics of hazardous waste in Los Angeles County? *Social Science Quarterly,* 78(4), 793–809.

44. Center for Effective Government. (2016, January). *Living in the shadow of danger: Poverty, race and unequal chemical facility hazards.* http://www.foreffectivegov.org/shadow-of-danger

45. Jerrim, J., & Macmillan, L. (2015). Income inequality, intergenerational mobility, and the Great Gatsby curve: Is Education the Key? *Social Forces,* 94, 505–533.

46. *Gregg, P., Jonsson, J. O., Macmillan, L., & Mood, C. (2017). The role of education for intergenerational income mobility: A comparison of US, Great Britain and Sweden. Social Forces, 96(1), 121–152.*

47. Reeves, R. V., & Fall, C. (2022, August 2). *Seven key takeaways from Chetty's new research on friendship and economic mobility.* Brookings. https://www.brookings.edu/blog/up-front/2022/08/02/7-key-takeaways-from-chettys-new-research-on-friendship-and-economic-mobility/

48. Hardin, D. J. (2009). Collateral consequences of violence in disadvantaged neighborhoods. *Social Forces,* 88(2), 757–84.

49. Wallace, D. (2012) Examining fear and stress as mediators between disorder perceptions and personal health, depression, and anxiety. *Social Science Research,* 41(6), 1515–1528.

50. Truong, K. D., & Ma, S. (2006). A systematic review of relations between neighborhoods and mental health. *Journal of Mental Health Policy and Economics,* 9(3), 137–154.

51. Curry, A., Latkin, C., Davey-Rothwell, M. (2008). Pathways to depression: The impact of neighborhood violent crime on inner-city residents in Baltimore, Maryland, USA. *Social Science and Medicine,* 67(1), 23–30.

52. WHO. (2021, December 9). *State of inequality: HIV, tuberculosis and malaria.* https://www.who.int/publications/i/item/9789240039445

53. WHO. (2015, May 5). *Monitoring health inequality: An essential step for achieving health equity.* https://www.who.int/data/inequality-monitor/publications/videos

54. Fuertes, C. V., et al. (2022). The association between childhood immunization and gender inequality. *Vaccines,* 10(7), 1032.

55. Richardson, E. T., Collins, S. E., Kung, T., Jones, J. H., Tram, K. H., Boggiano, V. L., Bekker, L. G., Zolopa, A. R. (2014). Gender inequality and HIV transmission: A global analysis. J. Int. AIDS Soc., 17(1), 19035.

56. LeGates, R., & Stout, F. (Eds.). (2004). Early Urban Planning V1 (1st ed.). Routledge. https://doi.org/10.4324/9781003101048

57. Simon, S. (2004). Industrialization and health. *British Medical Bulletin,* 69(1), 75–86.

58. CDC. (n.d.). *Health equity.* https://www.cdc.gov/chronicdisease/healthequity/index.htm

59. Murphy, S. L., Kochanek, K. D., Xu, J., & Arias, E. (2021). Mortality in the United States, 2020. NCHS data brief, (427), 1–8.

60. WHO. (2020, December 9). *The top 10 causes of death.* https://www.who.int/news-room/fact-sheets/detail/the-top-10-causes-of-death

61. WHO. (2014, December 30). *Global nutrition targets 2025: Low birth weight (LBW) (Policy brief).* https://www.who.int/publications/i/item/WHO-NMH-NHD-14.5

62. Risnes, K. R., et al. (2011). Birthweight and mortality in adulthood: A systematic review and metaanalysis.

Int J Epidemiol, 40, 647–661. https://doi.org/10.1093/ije/dyq267

63. CDC. (2022) Births: Final data for 2020. *National Vital Statistics Report, 70*(17). https://www.cdc.gov/nchs/data/nvsr/nvsr70/nvsr70-17.pdf

64. Congress for New Urbanism. (n.d.). *The charter of new urbanism* https://www.cnu.org/who-we-are/charter-new-urbanism

65. Cabrera, J. F. (2013). Can new urbanism create diverse communities? *Journal of Planning Education and Research, 33*(4), 427–441.

66. Brown, B. B., & Cropper, V. L. (2001). New urban and standard suburban subdivisions: Evaluating psychological and social goals. *Journal of the American Planning Association, 67,* 402–419.

67. Hill, L., & Artiga, S. (2022, August 22). *COVID-19 cases and deaths by race/ethnicity: Current data and changes over time.* Kaiser Family Foundation. https://www.kff.org/coronavirus-covid-19/issue-brief/covid-19-cases-and-deaths-by-race-ethnicity-current-data-and-changes-over-time/

68. CDC. (2022 June 15). *Science brief: Evidence use to update the list of underlying medical conditions associated with higher risk for severe COVID-19.* https://pubmed.ncbi.nlm.nih.gov/34009770/

69. Morens, D. M., Breman, J. G., Calisher, C. H., Doherty, P. C., Hahn, B. H., Keusch, G. T., Kramer, L. D., LeDuc, J. W., Monath, T. P., & Taubenberger, J. K. (2020). The origin of COVID-19 and why it matters. *Am J Trop Med Hyg, 103*(3), 955–959.

70. United Nations. (2021, September 28). *The magnitude and scope of inequalities created and exacerbated by COVID-19. https://www.ohchr.org/en/2021/09/magnitude-and-scope-inequalities-created-and-exacerbated-covid-19-truly-shocking-high*

71. Centers for Disease Control and Prevention. (2013). Health disparities and inequalities report. *MMWR, 62*(3). https://www.cdc.gov/minorityhealth/chdireport.html

72. Schuch, F. B., Vancampfort, D., Firth, J., Rosenbaum, S., Ward, P. B., Silva, E. S., et al. (2018). Physical activity and incident depression: A meta-analysis of prospective cohort studies. Am J Psychiatry, 175, 631–648.

73. Choi, K. W., Zheutlin, A. B., Karlson, R. A., Wang, M-J., Dunn, E. C., Stein, M. B., et al. (2020). Physical activity offsets genetic risk for incident depression assessed via electronic health records in a biobank cohort study. Depress Anxiety, 37, 106–114.

74. Ekelund, U., Tarp, J., Steene-Johannessen, J., Hansen, B. H., Jefferis, B., Fagerland, M. W., et al. (2019). Dose-response associations between accelerometry measured physical activity and sedentary time and all cause mortality: systematic review and harmonised meta-analysis. BMJ, 366, l4570.

75. Pickens, C. M., Pierannunzi, C., Garvin, W., and Town, M. (2015). Surveillance for Certain Health Behaviors and Conditions Among States and Selected Local Areas — Behavioral Risk Factor Surveillance System, United States. MMWR Surveill Summ 2018, 67(SS-9), 1–90. http://dx.doi.org/10.15585/mmwr.ss6709a1external

76. Robert Wood Johnson Foundation. (2023). State of childhood obesity. https://stateofchildhoodobesity.org/demographic-data/ages-10-17/

77. Gostin, L. O., & Gronvall, G. J. (2023). The origins of Covid-19: Why it matters (and why it doesn't). N Engl. J Med., 388, 2305–2308. https://www.nejm.org/doi/full/10.1056/NEJMp2305081

Chapter 9: What Can You Do?

1. https://www.census.gov/library/stories/2023/01/volunteering-and-civic-life-in-america.html
2. Ross, M and Sicoly F (1979) Egocentric biases in availability and attribution. Journal of Personality and Social Psychology, 37, 322-336
3. Pronin E, Lin D and Ross L (2002) The Bias Blind Spot: Perceptions of Bias is Self vs Others. Personality and Social Psychology Bulletin, 28, 369-381
4. Tocqueville, Alexis (1835) Democracy in America
5. Putnam, Robert (1995) Bowling Alone: America's Declining Social Capita. Journal of Democracy.
6. Sawhill IV. (2020) Social Capital: Why we need it and How we can create more of it. Brookings Institution. https://www.brookings.edu/wp-content/uploads/2020/07/Sawhill_Social-Capital_Final_07.16.2020.pdf
7. Pew Research, https://www.pewresearch.org/short-reads/2020/01/09/70-of-americans-say-u-s-economic-system-unfairly-favors-the-powerful/
8. Wall Street Journal Poll, https://nypost.com/2023/03/27/values-like-patriotism-religion-falling-out-of-favor-among-americans-poll/
9. https://www.cambridge.org/core/journals/american-political-science-review/article/is-there-a-culture-war-conflicting-value-structures-in-american-public-opinion/141B22BC55DFAB44ED65F045942AC3B1
10. Haidt, Johnathan (2006) The Happiness Hypothesis. Basic Books, NY, NY
11. Ardelt, M (2000) Still Stable after all these Years? Personality Stability Theory Revisited. State of Social Psychology, 63(4): 392-405
12. Emmons, RA (2003). Personal Goals, life meaning and virtue: Wellsprings of a positive Life. In. CLM Keys and J Haidt (Eds), Flourishing: Positive Psychology and Life Well Lived. (105-128). Washington DC, APA
13. Nakamura, J., & Csikszentmihalyi, M. (2003). The construction of meaning through vital engagement. In C. L. M. Keyes, & J. Haidt (Eds.), Flourishing; Positive psychology and the life well lived (pp. 83–104). Washington: American Psychological Association
14. Ignjatovic C, Kern ML, Oades LG (2022) The VIVA Sustainble Work Engagement Model, In J Appl Psychology, 7:251-270
15. Linley, P. A., Nielson, K. M., Wood, A. M., Gillet, R., & Biswas-Diener, R. (2010). Using signature strengths in pursuit of goals: Efects on goal progress, need satisfaction, and well-being, and implications for coaching psychologists. International Coaching Psychology Review, 5, 8–17
16. Mijares et al (1990) Mentoring: A Concept Analysis. The Journal of Theory Construction & Testing, 17(1): 23-28
17. United Nations (ND). IPCC. https://www.ipcc.ch/ (Accessed 1-23-23)
18. APA, Mentoring, https://www.apa.org/education-career/grad/mentoring
19. Pew Research (2021). Social Media use in 2021. https://www.pewresearch.org/internet/2021/04/07/social-media-use-in-2021/ Accessed 8-7-23
20. Brookings Institution (2019) American Values Survey. https://www.pewresearch.org/politics/values-questions/ (Accessed 5-4- 23)
21. Sandel M (2009) Justice. Farrar, Strauss and Giroux, NY, NY
22. Putnam, Robert (2001) Bowling Alone. Simon and Schuster. London, England
23. Graeber, David and Wengrow, David (2021). Dawn of Everything: A New History of Humanity. Farrar,

Strauss and Giroux, NY, NY

24. Christakis, Nicholas (2019). Blueprint:The Evolutionary Origins of a Good Society. Little, Brown Spark.

Glossary

Globalization	The economic, social, and cultural integration of the world. Economic globalization is the result of trade between regions and countries. Social and cultural integration is facilitated by trade, travel, and the internet.
Absolute advantage	Countries should specialize in all goods they can produce efficiently due to advantages in labor, technology, resources, or other favorable local conditions.
Active transportation	The movement of people and goods using nonmotorized means such as walking and bicycling
Age structure	The number of people in each age (one, two, three, etc.) or age range (0–4, 5–9, etc.) by sex (M, F). Age structure is sometimes called a "population pyramid."
Aggregate demand	The total spending by households, businesses, and government
Anthropogenic	Environmental impacts related to human activity
Area median income (AMI)	Median household income of a particular region. AMI provides a better measure of the central tendency of the area's income as it's not impacted by outliers as is average or mean income
Animal source foods	Increasing consumption of ASFs is implicated in the increase in GHGs, global warming, and numerous health problems related to obesity.
Autarky	An economic philosophy that stresses complete self-reliance and economic independence
Balance of payment	A summary of all economic transactions between a region's residents and nonresidents in a certain time period. Typically includes three accounts: current, capital, and financial. Ideally regions should have a zero balance of payment, which indicates a healthy economy (as exports and imports are equivalent). Countries that import more will have a negative BOP account—a large deficit—and may suffer economic consequences such as capital flight or devaluation of currency.
Biodiversity	Variation in lifeforms in an ecosystem. Can be measured in terms of the variation in genotypes, phenotypes, or species. Earth is estimated to have nearly 9 million species of plants and animals.

Biomedical	Physiological or pathophysiological characteristics of disease. The advent of germ theory in the 19th century helped spur the transition to a biomedical view of disease and its progression.
Body Mass Index	Wt in kilograms divided by height in meters squared
Bonds	Debt instruments sold by governments, corporations, and other entities to raise capital. Bonds are rated, as some are riskier than others. Usually higher risk bonds provide higher interest rates. The U.S. government bonds are also called T-Bills.
Bonds	Human health impacts related to insufficient food intake; can include macronutrient (protein-energy malnutrition) and micronutrient (vitamins, minerals) deficiencies.
Capital	Usually refers to money and the other things (e.g., machines) needed to make goods. Sometimes used to refer to those who control capital (i.e., the rich, factory owners)
Capital gains	The tax levied on profit.
Capitalism	An economic system in which scarce resources for goods and services are allocated based on supply and demand in an open market. Capitalism allows for private property ownership and assumes that people are self-interested, profit seeking, and likely to engage in activities that improve their financial standing.
Carbon dioxide	An important atmospheric gas that is used as the baseline to assess the GWP of other gases. Produced by the consumption of fossil fuels.
Chronic diseases	Diseases that are usually linked to lifestyle choices (e.g., food consumption, low physical activity) and/or environmental exposure and cannot be cured but must be managed to reduce morbidity and mortality. Chronic diseases are the biggest cause of mortality worldwide. Examples include heart disease, cancers, strokes, diabetes, and COPD.
Civic engagement	Becoming involved in matters outside your household to impact public concerns. Important for the sustenance of democracy.
Climate change	The various environmental impacts such as extreme weather associated with global warming
Cohort	A group identified as such on the basis of a shared characteristic (e.g., age cohort)
Collateralized debt obligations (CDO)	Consumer loans for homes, automobiles, and other products that are bundled together and sold to investors as a financial product
Colonization	The military, political, economic, and social control of one region by another. Colonization, in common usage, refers to the conquest and control of regions in Africa, Americas, and Asia by European states.
Communicable diseases	Diseases that are transmitted from person to person (e.g., COVID), through a vector such as a mosquito (e.g., malaria) or from the environment (e.g., parasites)

Comparative advantage	Countries should specialize only in those goods and services for which they have the lowest opportunity cost and engage in trade for other goods and services for which others have a comparative advantage.
Disability adjusted life years (DALYs)	Measure of years lost due to infirmity, disease, or premature death
Delation	The fall in prices of goods and services over time
Demand-side economic policies	The idea that demand of goods and services is the central driver of the economy. Demand-side policies seek to increase aggregate demand during economic slumps by making borrowing easier (e.g., cutting interest rates) or increasing government spending. John Maynard Keynes was a proponent of demand-side policies.
Dependency ratio	The number of dependents, or those below 15 and over 65, divided by those of working age (between 15 and 65). The dependency ratio provides insight into the burden imposed on those who are working by children or the elderly. In some countries, like Japan, falling fertility and high life expectancy has increase old-age dependency, which means there are increasingly fewer people paying into pensions and health care schemes to care for the elderly.
Dependency theory	The idea that developed nations need to exploit poorer nations and keep them in poverty to maintain their economic ascendancy
Deregulation	The reduction of government intervention, including regulation, oversight, and mandates, in markets
Distortions	Actions that interfere with the functioning of markets and impact supplier and consumer behavior and, by extension, price. Government regulation is a source of distortion.
Domains of action	These include the micro, interpersonal, meso, regional, and macro domains. These distinctions are more conceptual than actual. However, they can help guide where to best focus your energies given your abilities, proclivities (likes and dislikes), and constraints.
Empirical	Collection of data to test hypotheses.
Energy	Essential for work in all aspects of civilization. Produced by both easily replenished renewable sources such as wind or solar that have fewer environmental impacts and nonrenewable sources such as fossil fuels (coal, petroleum, natural gas) with significantly larger environmental consequences
Energy	Needed to do work (force over distance). Energy can be renewable (solar, wind, geothermal) or nonrenewable (coal, natural gas, petroleum). In contrast to nonrenewable energy, renewable sources are easily replenished and don't contribute to climate change.
Environmental Protection Agency	The federal department tasked with environmental regulation, particularly the NEPA, Clear Water, and Clean Air Acts.

Epidemiology	Epidemiologists are interested in describing the characteristics of a disease with respect to its origins, incidence, spread, prevalence, treatment and containment, and group differences in how the disease is experienced.
Epistemology	The process used to gain knowledge (or way of understanding phenomena)
Exchange value	The price of a good or service in a market (contrasted with exchange value)
Exposure	Presence of vulnerable populations
Externalities	An impact on entities outside the market. Though externalities can be good or bad, the word is often used pejoratively. The impact on ocean life due to plastic debris is an externality.
Factor accumulation	An increase in labor, capital, land, or entrepreneurship, the four pillars of production
Fauna	All animal life in a region
Fecundity	The biological capacity for reproduction
Federal funds rate	Rate paid on reserve balances at the Federal Reserve. This is a benchmark rate for other retail rates (e.g., credit cards) and interbank borrow rates (rates at which banks borrow from each other)
Federal Reserve	The institution tasked with managing the nation's money supply (monetary policy) and regulating the banking sector
Fiat currency	Currency that is not backed by gold or some other limited commodity. This allows nations to print as much money as they want (often creating inflation)
Fiscal policy	Actions taken the government (e.g., tax credits, tax deductions) to stimulate consumer spending.
Flora	All plant life in a region
Food deserts	Areas without access to fresh fruits and vegetables and stores sell mostly processed foods high in sugar, fats, and other harmful additives
Fossil fuels	Energy sources such as petroleum or natural gas formed by the decomposition of long dead animals and plants
Fractional Reserve banking	The legal ability of retail banks to use a portion of deposited money for loans and credit, thereby increasing the money supply in the economy.
Gross domestic product (GDP)	A measure of economic activity in a region; the sum total of the value of the final goods and services made in a region.
General plan	A plan guiding the development and land use decisions in a city within the context of certain elements like housing, circulation (transportation), safety, noise, conservation, and environmental justice.

GINI	An index used to measure inequality. A score of 0 indicates perfect equality and 1 is perfect inequality.
Glocalization	Production is located in another country while corporate HQ and R&D are still in the country of origin
Green revolution	The use of fertilizers, genetic modifications, and other technologies to increase agricultural yield
Greenhouse gases	The gases in the atmosphere (carbon dioxide, methane, nitrous oxide) that trap heat, causing global warming, an increase in the planet's average temperature
Global warming potential (GWP)	Each gas has a GWP that is measured relative to CO2, which has a GWP of 1. The higher the GWP, the greater the impact on global warming.
Haber–Bosch process	A method created by Fritz Haber for extracting nitrogen, a vital component of fertilizer, from the atmosphere. This process has, essentially, fed the world, creating the population boom of the 20th century.
Hazards	Potential for the occurrence of events that have the potential for negative impacts
Headship rate	The probability that a person in a certain age range is likely to be head of household
Health disparities	The disproportionate disease risk and harm experienced by certain groups due to social, biological, and other factors
Higher density development	More building units per area (acres, square feet). Planners often advocate for higher density to increase housing units to reduce housing cost; increase walking and bicycling as more units means more and different land uses (residential, business, recreation) will be located closer together; and reduce pollution (as travel times will go down).
Household and family	A household is a group of people, related or unrelated, living in a housing unit. In a family household, occupants are related by birth, marriage, or adoption.
Householder	The person in whose name the house is owned or rented
Human Development Index (HDI)	An indicator that combines measures of economic development (e.g., per–capita GDP) with other measures of well–being (education, life expectancy)
Immigration	The movement of people from one area to another. People leaving an area emigrate, while those coming into an area immigrate.
Import substitution (ISI)	Economic policy adopted by some nations that emphasized replacing foreign imports with local or indigenous production. In some cases, steps taken to privilege local producers created adverse consequences such as price fixing due to local monopolies and reduced product quality as foreign competition was missing.
Incidence	The number of new cases of a condition in a region in a particular time period. The incidence rate for a disease would be the number of people who contract the disease in a given time period divided by the total population at risk.

Income	Taxable wages paid for services or labor
Industrial revolution	The advent of mechanization in production starting in Europe/United States in the 1760s and continuing through the mid 1800s (e.g., factories, assembly lines) that reduced production time and increased product consistency and quality
Inequality	The unequal distribution of income and wealth or increasing concentration of income and wealth among fewer individuals or a group
Inflation	The increase in prices of goods and services over time
International Monetary Fund (IMF)	Created in the aftermath of WWII to stabilize trade and the world economy by providing loans and economic advice, particularly to emerging economies
Intergovernmental Panel on Climate Change (IPCC)	Convenes experts to assess current research on climate change. Findings are reported in assessment reports (five as of 2022).
Knowledge, attiudes, and practices (KAP)	Framework used to inculcate healthier behaviors in public health
Labor	Those providing work for wages paid by capitalists to turn unfinished raw materials into finished goods and services for sale in a market
Landfills	Waste is taken to landfills for storage and processing. Some landfills use the gas from decomposition to produce energy (called bioreactor landfills)
Leadership in Energy and Environmental Design (LEED)	A sustainability or green building rating system developed by the U.S. Green Building Council. There are several categories for which LEED certification is available, including new construction or neighborhood design. Ratings vary from bronze to platinum.
Life expectancy	The additional number of years a person can be expected to live at every age if the age-specific death rates for a given year prevailed for the rest of the person's life. Life expectancy at birth is the additional number of years a newborn can be expected to live. Women usually have longer life expectancies at all ages than men.
Low birth weight (LBW)	Birth weight that is below 2,500 grams, or 5.5 lbs. Babies below this threshold are at higher risk of significant short–term and long–term health problems. Preterm LBW babies tend to do much better than IUGR (intrauterine growth restriction) babies or full–term babies who are still LBW.
Macronutrients	Energy sources for the body. Essential for maintaining the structure and function of different parts of the body. Includes carbohydrates, protein, and fats.
Malnutrition	Human health impacts related to insufficient food intake, including macronutrient (protein-energy malnutrition) and micronutrient (vitamins, minerals) deficiencies.

Markets	A place, real or virtual, that brings together buyers and sellers. Markets should be transparent and free from distortions (e.g., monopolies or regulation) and coercion (buyers and sellers can enter and leave as they please). Markets assume that individuals are self–interested and rational. These characteristics enable perfect competition.
Marxism	The political-economic ideas of Karl Marx and Friedrich Engels who saw history progressing from serfdom to capitalism and ultimately to socialism. Marxist ideas emphasize historical materialism (the impact of the economic system on social structures), class conflict, and dialecticism (thesis-antithesis-synthesis).
Means testing	The provision of welfare based on income and assets.
Medicaid	A government health care insurance program for the elderly
Medicalization	The process by which parts of the social world are brought under medical or expert control
Medicare	A government health care insurance program for poor adults and children
Mercantilism	An economic philosophy that emphasizes exports. Mercantilist nations extracted (i.e., took) resources from their colonies and used the resources to create finished goods, which were exported back to the colonies and others.
Methane	A common global warming gas produced by agriculture and waste processing, specifically enteric fermentation in cows. An increase in factory farming for beef has contributed to the rise in methane levels.
Micronutrients	Vitamins and minerals
Modernity	The current historical period characterized by the norms, values, and culture that derive from a focus on reason, positivism, empiricism, and the scientific method. Features include scientific and technological growth; an increase in individual freedoms and democracy; the rejection of older, nonrational belief systems; and reliance on market-based economies.
Monetary policy	Actions taken by the Central Bank (e.g., U.S. Federal Reserve) to increase or decrease the money supply.
Money	Any item that market participants have agreed on as a "unit of account" so that prices for different goods and services can be paid using whatever serves as money. Anything used as money should be both convenient to use and exchange.
Money supply	The total amount of money in an economy. Includes cash, coins, bank deposits, and other cash equivalents (e.g., short–term CD) that can be converted into money quickly. M1 includes cash and deposits, and M2 includes CDs, time deposits, or money market investment accounts.
Morality	Principles used to determine right from wrong. Examples include utilitarianism, Kantian categorical imperatives, religion (e.g., biblical principles). Moral beliefs are the foundation for the rules and laws used to organize society.

Morbidity	The impact of sickness measured in terms of intensity and duration of symptoms. Morbidity can be measured for diseases, injuries, and disabilities. Related concepts include disability-adjusted life years (DALY) and quality-adjusted life years (QALY)
Mortality	Statistics related to death. The crude mortality rate is measured by dividing the total dead by total population. Mortality rates can also be calculated for diseases (e.g., COVID mortality rate), age ranges, or any other socially meaningful category.
Municipal solid waste	Waste produced inside cities
New urbanism	A planning philosophy that stresses higher density development, mixed uses, inclusivity across various dimensions, human scale design, public spaces, social capital, active transportation, diverse economic activities, and integration with nature
Nitrous oxide	A common global warming gas produced by agriculture and wastewater treatment.
Nuptuality	The marriage rate by sex
Organization for Economic Cooperation and Development (OECD)	Includes 38, mostly higher income, countries that collectively account for 80% of the world's trade. Primarily an advisory body that convenes experts and policy-makers to foster dialogue, set standards (over 450 international standards have been developed through its activities), and formulate policy.
Ontology	The branch of philosophy concerned with questions of "being" or the kinds of things that are "real"
Opportunity cost	Indirect cost or unrealized gain for other paths or alternatives not taken (contrasted with direct cost or actual costs incurred in the production of something). Essential for understanding comparative advantage.
Overweight and obesity	High BMI, or Body Mass Index (weight (kg) / height (M) squared), due to excess consumption of foods, especially processed foods or foods that have been produced or refined artificially
Ozone	Composed of three oxygen atoms. Protects earth from ultraviolet radiation.
Pandemic	A countrywide or worldwide outbreak of disease
Per capita income	The GDP divided by population
Police power	The legal power of the government to regulate. In the United States, the power of the government to regulate health, safety, welfare, and morals is explicitly mentioned in the U.S. Constitution.

Poverty, poverty line, poverty rate	An income threshold below which individuals and households are considered low income. In developed countries, poverty is relative (or measured with respect to the incomes of higher earning groups). In developing countries, poverty is absolute (below \$2.15 per day) in that those below the poverty line are at serious risk of not having enough money to procure essentials for survival. The poverty rate is a measure of how what percentage of people are below the poverty line in a country or region.
Prevalence	The number of people who have a condition in a region at a certain point in time. The prevalence rate for a disease would be the number of people with the disease (i.e., the cured or dead are excluded) at a certain point divided by the total population at risk.
Price	What is paid for a good or service. Results from the intersection of supply and demand for a good or service. Usually prices increase when demand is greater relative to supply and decrease when the opposite is true.
Primary prevention	Measures taken to prevent certain health conditions from occurring. This is the cheapest and most effective way to address health problems.
Privatization	The provision of goods and services by the suppliers competing in a market (as opposed to state or government owned companies)
Pro-environmental behaviors (PEB)	Behaviors that impact the environment positively.
Property ownership	The legally granted right to possess, control, use, dispose, and prevent others from using
Purchasing power parity (PPP)	Adjustments made to the GDP to account for the differences in the costs of goods and services between regions
Quality-adjusted life years (QALYS)	Measure of quantity of life lived and its quality (1 QALY is equivalent to 1 year in perfect health)
Rate	A metric calculated by dividing one number by another to estimate the frequency of an event (e.g., death rate = number of deaths divided by total population)
Ratio	A metric that uses two numbers to estimate how many times one number contains the other to illustrate the relative strength of each number (e.g., sex ratios are the number of males divided by the number of females. In areas with selective abortions of females, you may see a sex ratio more skewed toward males. You can also take the ratio of rates, such as the ratio of death rate in one region to another to assess the relative risk of dying in one region versus another.)
Recycling	Reusing waste in some capacity to create new products, thereby reducing the need to store waste in landfills and the negative environmental impacts associated with waste production. This process varies greatly by material.
Reduce, reuse, recycle (RRR)	The tenets of sustainable waste management. The first priority is to reduce consumption and waste, the second is reusing items to increase their life span, and the third is recycling materials to divert them from the waste stream.

Remittances	Funds sent back to the country of origin by those working or living abroad (i.e., someone who is part of the diaspora)
Rents and rent–seeking behavior	Any wage or price that exceeds the market price for labor or a good. Any actions taken by the wealthy to protect their wealth and increase their wealth without increasing economic productivity.
Reserve currency	A currency, such as the U.S. dollar, held by other nations as part of their foreign exchange reserves because that currency is commonly accepted throughout the world for payment
Reserve ratio	The percentage of total bank deposits that must be held at the Federal Reserve
Resources	Anything that can be used for economic purposes; includes land, labor, capital, technology, and so on
Retail banks	An essential part of the modern economy as these banks hold consumer deposits, facilitate spending (e.g., checking accounts, credit cards), provide consumer loans (e.g., homes, cars), and help expand the money supply through fractional reserve banking, which allows them to lend a portion of the amount held as deposits. An example is the Bank of America or HSBC.
Risk	Potential for adverse consequences
Sample and population	A smaller subset of a larger group or population. When collecting data, researchers often rely on samples to draw conclusions about a larger population of interest.
Sustainable Development Goals (SDGs)	The UN's 17 goals addressing poverty, hunger, education, energy, water and sanitation, industry, health, climate change, urban development, work and labor, justice and equity, and flora/fauna.
Secondary care	Treating a disease or condition after it has occurred with the intention of curing it
Sequalae	Ongoing impacts of an acute or short–term conditions. COVID symptoms that persist after the initial disease has subsided is an example.
Social capital	The social networks, formal and informal, in a society that act as a social glue allowing for cooperation, solidarity, cultural coherence and the emergence of a community
Social construction	The idea that beliefs, values, and other important ideas about people, objects, and phenomena arise from social forces. For example, the American idea that government should help children or elderly because they are vulnerable and deserving and not needy working–age adults is a social construction.
Social determinants of health	Social factors such as SES, education, health care access, and neighborhood or community–related factors (e.g., crime, pollution) that impact disease exposure, response, and outcomes.
Social Security	U.S. federal income and benefits program for the elderly and disabled

Specialization	Usually refers to work specialization. This allows workers to become more efficient and, ultimately, productive. This also fosters interdependency as no one individual can produce everything they need to survive. This is the foundation of modern global capitalism.
Structural adjust-ment policies	Opening economies to outside competition along with reduced protectionism and government spending and regulation; used historically by the IMF for economic development
Structure and agency	Forces that are outside the individual and impact the individual are structural. The capacity of the individual to take action to shape their own destiny is agency. These perspectives are used to understanding social problems. For example, why are some poor or commit crime? Is it due to structural factors (i.e., neighborhood effects) or agency (i.e., choices made by the individual)?
Subsistence agriculture	Farming or the growing of food for one's family or household with little or no surplus for trade or sale
Supply chain	The different parts of a finished good (e.g., cars) are often manufactured in different places based on the comparative advantage of the location for that part. The supply chain is comprised of the various locations in which a product's constituent parts are made.
Supply–side eco-nomic policies	Economic theory that emphasizes the production or provision of goods and services as the engines of economic growth. Supply–side policies (e.g., Reaganomics) seek to cut business taxes or capital gains taxes for investors as a way of stimulating the economy.
Surveillance	A core function of public health; the monitoring of health indicators to identify new threats, their progression, and outcomes
Sustainability	A philosophy that emphasizes identifying and addressing important issues today so that the world—humans and other flora and fauna—thrive in the long term. Typically defined in terms of three broad parameters (economic, environmental, and equity). Economic sustainability includes poverty alleviation and development of economies that provide a high standard of living for all; environmental sustainability includes the use and recycling of resources that minimize pollution, waste, and externalities, or harms to those outside the economic system; equity refers to societies that are nondiscriminatory, inclusive, and participatory across various dimensions such as gender, sexual orientation, age, race, and so forth.
Tax credit	A "dollar–for–dollar" cancellation of tax liability used to stimulate spending. Electric vehicles or solar power have been incentivized in the United States through the use of tax credits. A $100 tax credit means $100 less in tax liability.
Tax deduction	Allowable reduction in taxable income due to some type of fiscal policy. Mortgage interest in the United States can be deducted from gross annual household income.
Taxes	An amount payable to government on specific economic transactions. Taxes are levied by all branches of government on income, real estate (property tax), sale of consumer goods (sales tax), and other items.

Teleology	The belief that in various domains (e.g., nature, economy, etc.) there are specific, desirable, outcomes; the rationality of actions is measured with respect to how likely the actions are to achieve those desirable ends. Democracy and global capitalism are sometimes framed in teleological terms (i.e., these outcomes are desirable and rational, and governments must act to achieve them). Teleological thinking can make it difficult to imagine other alternatives, adapt, and change.
Tenure	The arrangements that characterize how a household occupies a housing unit. Usually housing units are rented or owned, though other arrangements are also possible.
Tertiary care	Managing a disease or condition that cannot be cured. Most chronic diseases impacting developed countries and the wealthier populations of poorer countries require tertiary care. Examples include heart disease, hypertension, and diabetes.
Total fertility rate (TFR)	The average number of children a woman can be expected to have as she goes through her childbearing years, typically 15–44. The TFR provides a measure of the number of children women have on average. Replacement–level fertility, the number needed for a population to reproduce itself in the absence of immigration, is usually 2.1 children per woman.
Trade	The exchange of goods, services, and other items (e.g., cultural artifacts) between regions. Specialization and trade are the foundation of the modern global capitalism system.
Trickle–down theories	Idea that spending at the top income or wealth tiers eventually makes it way down to the bottom tiers (i.e., spending by the rich will eventually benefit all tiers of society including the poor). There is very limited empirical evidence for trickle–down effects.
Underemployment	The proportion of people taking work that is not full time and/or for lower wages because they were unable to find appropriate work
Underwriting	Evaluating the credit worthiness of a potential borrower
Unemployment	The proportion of workers seeking work who cannot find work
Universal basic income (UBI)	Income provided to all adults in a given region
Urbanization	A process that results in the formation of cities. Characteristics of urban areas include large populations, higher populations densities, increasing dependency on markets for goods and services, diversity (cultural and social), greater resource consumption, and waste.
U.S. Census and ACS	The census is an attempted count of all U.S. residents and important characteristics. The full Census occurs every 10 years. In the intervening years, the American Community Survey (ACS) samples a smaller population to generate generalizable demographic data for the nation.
Use value	The value derived from the use of a particular thing (e.g., house has an exchange value or price and an use value or shelter)

Usufuctory	The right to use and enjoy land (or another thing) without owning it
Utility	Choices made by individuals in a market to increase some measure of personal well-being
Vaccines	A substance (e.g., inactive viral particle, viral proteins) that stimulates full or partial immunity against a potential disease or as a therapy against an occurring disease
Vulnerability	The higher likelihood of being impacts negatively because of biological, social, or other factors
WASH	Inadequate water for various human needs, resulting in economic, social, and health–related impacts. Scarcity and stress are the result of increasing population and changing consumption patterns (e.g., more meat eating or fast fashion leads to greater water use), environmental factors (pollution, climate change), and under-developed water infrastructure.
Waste	Discarded materials from human activity. Recycling can help divert items from the waste stream, reduce environmental impacts, and reclaim a portion of the original economic value. Waste includes nonmunicipal waste due to mining, agriculture, and industry, which is often handled in situ (or where it was produced) and municipal solid waste (MSW), waste produced by various societal sectors (residential, commercial, industrial) that is taken to landfills, openly dumped, incinerated, or recycled.
Wealth	The value of assets (usually nontaxed until sold) such as real estate, stocks, precious metals, and even art
Weather	Daily atmospheric conditions such as temperature, humidity, rainfall, or wind
Welfare	Any type of government help based on legally defined eligibility criteria
World Bank	Created in the aftermath of WWII to provide loans, particularly to emerging economies, for infrastructure and economic development and poverty reduction
World Trade Organization (WTO)	The successor to the General Agreement for Trades and Tariffs (GATT, also created in the aftermath of WWII) to facilitate trade through its multilateral trading system and provide a venue for resolving trade disputes
Zoning	Land use management system used by cities to divide areas into zones, each with allowed and prohibited uses, design guidelines, and building standards
Zoonotic	Diseases passed from animals to humans